GLACIERS OF CALIFORNIA

CALIFORNIA NATURAL HISTORY GUIDES

Phyllis Faber and Robert Ornduff, General Editors

LATEST TITLES IN THE SERIES:

59. Glaciers of California: Modern and Ice Age Glaciers, the Origin of Yosemite Valley, and a Glacier Tour in the Sierra Nevada, by Bill Guyton
60. Sierra East: Edge of the Great Basin, edited by Genny Smith
61. Natural History of the Islands of California, by Allan A. Schoenherr, C. Robert Feldmeth, and Michael J. Emerson
62. Trees and Shrubs of California, by John D. Stuart and John O. Sawyer

CALIFORNIA NATURAL HISTORY GUIDES, 59

GLACIERS OF CALIFORNIA

Modern Glaciers,
Ice Age Glaciers,
Origin of Yosemite Valley,
and a Glacier Tour
in the Sierra Nevada

BILL GUYTON

UNIVERSITY OF CALIFORNIA PRESS

Berkeley Los Angeles London

University of California Press
Berkeley and Los Angeles, California

University of California Press, Ltd.
London, England

First Paperback Printing 2000

Library of Congress Cataloging-in-Publication Data

Guyton, Bill, 1932–
Glaciers of California : modern glaciers, ice age glaciers, origin of Yosemite Valley, and a glacier tour in the Sierra Nevada / Bill Guyton.
p. cm. —(California natural history guides ; 59)
Includes bibliographical references and index.
ISBN 0-520-22683-6 (pbk. : alk. paper)
1. Glaciers—California. I. Title. II. Series.
GB2425.C2G89 1998
551.31′2′09794—dc 21 97-44790
CIP

Printed in the United States of America

09 08 07 06 05 04 03 02 01 00
9 8 7 6 5 4 3 2 1

The paper used in this publication meets the minimum requirements of ANSI/NISO Z39.48-1992 (R 1997) (*Permanence of Paper*). ♾

CONTENTS

ILLUSTRATIONS

FIGURES

PLATES *following page 72*

TABLES

PREFACE

Glaciers are unusual and beautiful natural features whose presence make mountains more interesting, and, in the Ice Age, glaciers created attractive mountain scenery in places where no glaciers exist today. For these reasons glaciers and the Ice Age are naturally interesting subjects to many people, especially those who enjoy fishing, hiking, and climbing in California's mountains. What might be termed "glacio-tourism" is big business in California, as was shown by the economic distress and public outcry accompanying the temporary closure of Yosemite National Park in late 1995 and again in early 1997. Travelers, hikers, naturalists, backpackers, and photographers, many from far away, want to understand the glacial scenery of California, and Californians employed in education and the tourist and outdoor recreation industries should know about California's glacier story.

To prepare this book I have compiled, organized, and condensed information from several hundred sources published in numerous scientific journals, books, magazines, and government documents during the last 133 years. Most of these publications are out of print and/or are hard to find except by persons familiar with the geologic literature and who have access to a university library.

Curiosity and interest are the only prerequisites for reading and enjoying *Glaciers of California*. Special features of the book include a field-trip guide for those who want to see glaciers and glacial landforms firsthand, a glossary, quotes from original sources, and many references to help those who wish to learn more about specific areas or topics. Published articles and books are identified in the text by the author's last name and year of publication (for example "Jones, 1992"); the complete reference for each is at the end of the chapter.

It takes many people to "write" a book. Thanks to the following persons for help in various ways: Bill Babb, Cathy Brooks, Bill Jones, Tom Kearney, Ariaratnam Lakshmi, and Al Shnayer. Also, thanks to those who contributed photographs and drawings, who are named individually in the text. Special thanks are due to my wife, Dene, for assistance in photography and manuscript preparation; Doris Kretschmer and Arthur Smith at University of California Press for guidance and encouragement; Douglas H. Clark at Indiana University-Purdue University at Indianapolis for carefully reading two drafts and giving valuable suggestions and criticisms; D. D. Trent at Citrus College, who helped in numerous ways; Mary Hill, formerly of the California Division of Mines and Geology and the U.S. Geological Survey, and Howard Stensrud at California State University, Chico, for reading all or part of the manuscript and making suggestions about content and style; N. King Huber of the U.S. Geological Survey for helping me understand Yosemite Valley and the work of François Matthes; Elise Mattison of the California Division of Mines and Geology for help in preliminary work; Bonita Hurd for final copyediting; and California State University at Chico, and the Department of Geosciences, for bibliographic and material support. Uncredited photos throughout the book are by the author.

This book is dedicated to everyone who loves the High Sierra.

PART 1

CALIFORNIA AND GLACIERS

CHAPTER ONE

CALIFORNIA, GLACIERS, AND THE ICE AGE

GLACIERS IN CALIFORNIA?

Mention glaciers and few people will think of California. The Alps, the Canadian Rockies, or Alaska, yes—but not sunny California. However, California has several hundred small glaciers and glacierets in its high mountains, as well as beautiful scenery created by large glaciers during the Ice Age. Figure 1 shows areas of California that have been glaciated. The glaciated portion of the Sierra Nevada, the High Sierra, is one of the most beautiful and accessible mountain playgrounds in the world, and it is visited by large numbers of tourists, hikers, and backpackers each summer, many from other states and nations. In addition to providing inspiration and recreation, glacial California is a fine natural-science laboratory, and studies in California mountains have contributed much to our knowledge of glaciers and the Ice Age.

For glaciers to flourish an area needs snow, and California gets plenty of it. Abundant snow falls on the Klamath Mountains and Cascade Range because of their northerly location and proximity to the ocean. California's largest mountain range, the Sierra Nevada, is especially well situated for glaciers. Its 400-mile north-south extent, nearly perpendicular to the west-to-east storm track, plus its great height, which averages almost two miles in elevation, creates abundant orographic winter snow as moist air from the Pacific Ocean moves up and over the range. One recording station, Tamarack, located at an elevation of 8,000 feet, received an average of 38 feet of snow a year over a 22-year period, about forty times the snowfall of Antarctica, location of the world's largest glaciers. In one exceptional year, 73 feet of snow fell at Tamarack. The gentle western slope of the Sierra Nevada provides

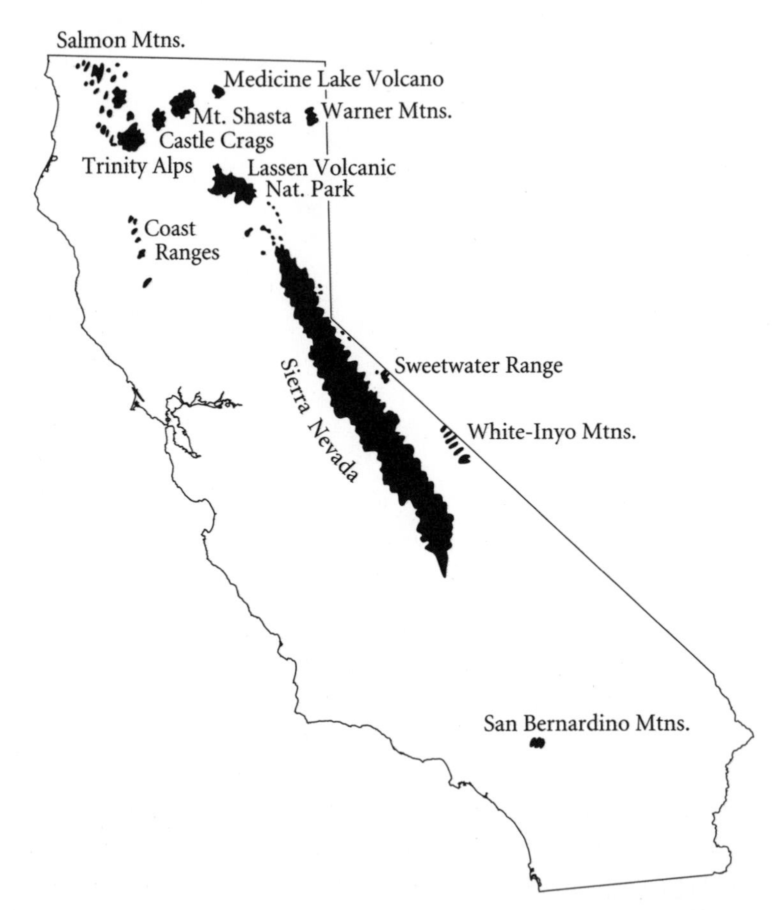

Figure 1. Areas in California that were glaciated during the Ice Age. Small glaciers exist today at Mount Shasta, in the Sierra Nevada, and in the Trinity Alps. (After Wahrhaftig and Birman, 1965.)

ample space for snow to accumulate. In the southern part of the range high elevation compensates for low latitude, while in the northern part high latitude compensates for low elevation.

If California mountains receive abundant snowfall today, why are their glaciers small compared to those of 20,000 years ago, when glaciers up to 60 miles long existed? Snowfall is only part of the recipe for making a glacier. Snow must be preserved so that over a period of years the amount of snow accumulated exceeds snowmelt. Antarctica gets little new snow each year, but virtually all of it is preserved to become glaciers because not much melting occurs in the summer. California's sunny, warm summers, prized by campers, backpackers, and chambers of commerce, cancel winter's fine effort and leave little snow to feed our glaciers. It is interesting to note in this regard that a rather modest increase in summer cloudiness might be all that is needed for California's glaciers to become large once again.

WHAT IS A GLACIER?

A glacier is a mass of ice formed from snow, on land, that exists over a period of years and is of sufficient thickness to flow downhill as a result of its own weight. A frozen lake, river, or waterfall is not a glacier. A frozen ocean, as in the Arctic, is not a glacier. A snowbank that persists for years is not a glacier but might become one if it becomes big enough and thick enough for flow to occur.

That ice can flow is contrary to everyday experience. An ice cube from the freezer placed on a table does not flow, it just sits there and melts. It is too small for flow to occur. An ice cube thousands of times larger would sag under its own weight and flow away as a glacier. Flow in ice is caused by gravity, the same force that makes the water of a river flow downhill. When gravity-induced stress exceeds the force that holds the particles of water or ice together, the atoms and molecules slip past each other downslope. To exceed ice's internal strength the downslope force produced by gravity must exceed some minimum value, and this is why an ice cube cannot flow. But as size increases, gravity-induced stress becomes greater while internal strength remains the same. Thus at some minimum thickness flow will begin in ice, or in any material for that matter.

Glaciers are often referred to as "rivers of ice" even though ice flows much slower than water, perhaps less than a foot a day. Ice has a greater internal strength than water so there must be more gravity-induced stress for flow to begin. A mass of ice will begin to flow when it is about 120 feet thick, a little less on a steep slope or in a warm climate, a little more on a gentle slope or in a cold climate.

Flow of ice caused by internal deformation, just discussed, is called *creep.* There is another type of glacier flow called *slide.* Slide occurs between the bottom of a glacier and the underlying rock and is aided by water at the base of the ice. Usually there is also mud, sand, or large pieces of rock in the ice at the bottom of the glacier, and slide causes this hard material to abrade and smooth the underlying bedrock, producing glacial polish and striations, as will be discussed and illustrated elsewhere.

Glaciers serve as valuable, natural water storage reservoirs (fig. 2) without any harmful environmental side effects such as those dams sometimes create. As Oliver Kehrlein wrote,

Figure 2. Glaciers on Sawtooth Ridge supply water to pastoral Bridgeport Valley, far below, during summer, fall, and drought years.

> California could expect a drought every summer . . . were it not for nature's method of equalizing excess runoff and damming it back at the source with mountain snow-packs and glaciers. This is one way she has of providing for her lowland flora and fauna during the arid season. (Kehrlein, 1948, p. 18)

To increase water supply during summer, some glaciers in China have been mined by spreading a black powder, such as coal dust, over the ice. The dark dust increases absorption of solar radiation and increases melting. Like all mining, this cannot be done forever.

HOW AND WHERE GLACIERS FORM

Glaciers cover 10 percent of the land on Earth today, and they form wherever snowfall exceeds snowmelt over a period of years. Conditions suitable for glaciers exist at high elevation, at high latitude, and at favorable combinations of elevation and latitude. Glaciers exist at sea level in Alaska and Antarctica and on especially high mountains even at the equator. In midlatitudes glaciers exist at in-between elevations, lower in British Columbia than in Washington, lower in Washington than in California, and lower in California than in Mexico. Coastal mountains usually have larger glaciers than areas far from the ocean, because snowfall is preceded by evaporation of water into the atmosphere. This is why California during the Ice Age had glaciers as much as 60 miles long when parts of central Alaska and Siberia, far from the ocean, had no glaciers at all.

To illustrate how glaciers form, imagine a place where, at the beginning of each winter, one foot of snow remains from the preceding winter. In a century there will be 100 feet of snow, a good start toward forming a glacier. But we must allow for compaction and recrystallization of snow into ice. Glacier ice is much denser than snow, and it may take 1,200 years and 1,200 feet of excess snow to make 120 feet of ice, a thickness sufficient for flow to begin.

As each layer of snow is compressed by the weight of overlying layers, the snow becomes granular, air is forced out, ice crystals grow together, density increases, and snow gradually becomes ice. The intermediate stage between snow and ice is called *firn* (a German word meaning "old snow") or *névé* (French). The change in specific gravity during this conversion is from about 0.1 for snow to 0.9 for ice (water is 1.0).

KINDS OF GLACIERS

One way of describing glaciers is by their temperature. Those in low or moderate latitudes are usually *temperate glaciers,* or *warm glaciers,* because almost all of the ice of the glacier is at its melting point the year around. Other glaciers, usually at high latitude or very high elevation, have ice that is, for the most part, colder than its melting point, and these glaciers are called *polar glaciers,* or *cold glaciers.* Compared to cold glaciers, warm glaciers have more water on and in the ice, release more water in the summer, flow faster, and erode more effectively. California, appropriately, has warm glaciers today, and Ice Age glaciers were probably warm as well.

Another way of classifying glaciers has to do with shape and size. *Confined glaciers,* or *mountain glaciers,* are of two types: *valley glaciers* and *cirque glaciers. Valley glaciers* (fig. 3) form in mountains and are confined to a valley, commonly one eroded previously by a river. Like rivers, valley glaciers are long and narrow, and they may join together downstream to form a trunk glacier. A representative valley glacier is a few miles long, a half mile wide, and a thousand feet thick. There are many thousands of valley glaciers in Alaska, Canada, and the Alps. The longest valley glacier in Alaska is over a hundred miles long.

A *cirque glacier* (figs. 3 and 4), smallest of all glaciers, is located at the head of a glacial valley in a steep-walled amphitheater called a cirque (rhymes with "jerk"). A representative cirque glacier might be half a mile long, half a mile wide, and a few hundred feet thick. If a valley glacier is a river of ice, a cirque glacier is a pond of ice. If a cirque glacier grew larger it would become

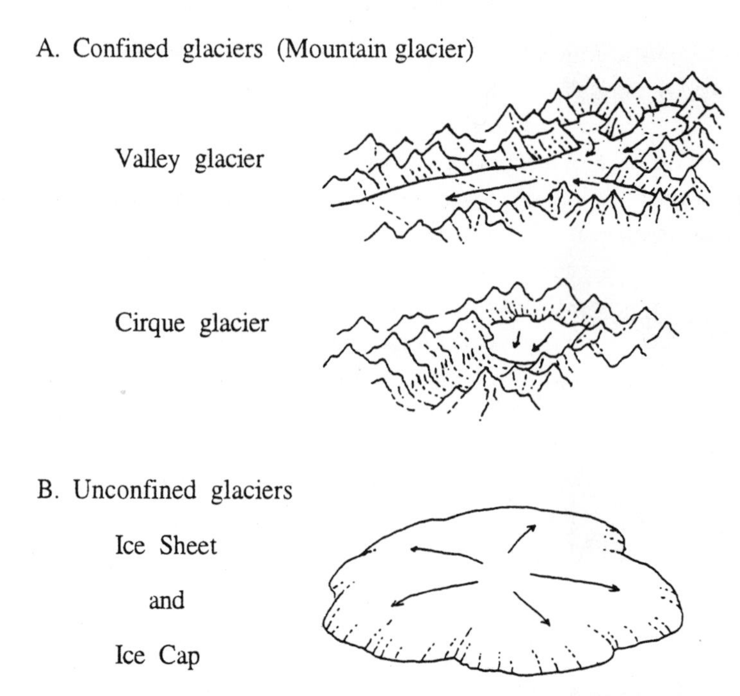

Figure 3. Types of glaciers. A valley glacier is long, like a river. A cirque glacier is small and only occupies the valley head. An ice sheet has an area greater than 50,000 square kilometers, while an ice cap is smaller. (Drawing by Susan Van Horn.)

a valley glacier. Sometimes it is difficult to tell if a permanent mass of snow and ice is a glacier or not, so the word *glacieret* is used to indicate small size and uncertainty as to whether or not the ice is flowing.

Unconfined glaciers are not constrained by mountains or former river valleys: the ice completely buries the topography and is free to flow in several directions, as syrup spilled on a table does. These largest of all glaciers can be hundreds of miles in diameter, be millions of square miles in area, and have ice two miles thick. An unconfined glacier is either an *ice sheet* or an *ice cap* (fig. 3). An ice sheet may form in mountains or on plains; if it occurs in mountains the ice flows over divides and completely buries all but the highest peaks. The Antarctic ice sheet is about 5 million square miles in area, yet it is smaller than the great ice sheet that once covered much of North America. Because of their great size, ice sheets are often referred to as *continental glaciers.* An ice cap is similar to an ice sheet but smaller (officially less than 50,000 square kilometers in area). To complete the analogy, if a valley glacier is a river of ice and a cirque glacier is a pond of ice, then an ice cap is a sea of ice and an ice sheet is an ocean of ice.

Figure 4. The Palisade Glacier, largest in the Sierra Nevada, is a cirque glacier. This photograph, taken August 24, 1972, shows snow in the zone of accumulation, bare ice in the zone of wastage, and the firn line separating the two zones. Dark coloration of the ice is a result of abundant rock debris in and on the ice. Prominent moraines surround the ice. The meltwater pond at the terminus of the glacier suggests the moraine has an ice core that prevents water from seeping through. The prominent lobes of the moraine may be caused by flowing glacier ice under the moraine. (Photo by Austin Post, U.S. Geological Survey.)

A *highland ice field,* or simply an *ice field,* is intermediate between confined and unconfined glaciers. An ice field forms where ice overfills valleys and merges into a common surface. The ice at depth flows confined, but the highest ice can flow across drainage divides. A highland ice field is smaller than an ice cap and does not have a convex surface. Highland ice fields were present in the Sierra Nevada during the Ice Age and were probably more common in California than has been recognized until recently (Clark, Clark, and Gillespie, 1991).

To understand these glacier names imagine a continuous sequence from small to large. Start with a snowbank on a mountainside. With favorable climate and increasing size it might become a perennial snowbank, then, in turn, a glacieret, a cirque glacier, a valley glacier, a highland ice field, an ice cap, and finally an ice sheet.

California today has 1 valley glacier, 107 cirque glaciers, and 401 glacierets. Ice Age California had numerous highland ice fields, hundreds of valley glaciers, and many more cirque glaciers. Some writers have in the past used the name *ice cap* in California, but this usage differs from today's definition, and the name ice field should be substituted for ice cap when reading the older publications. California does not appear ever to have had a true ice cap or ice sheet.

CREVASSES

Crevasses are open cracks, or fissures, in the ice of a glacier that are several feet wide and hundreds of feet long. Crevasses form because the ice in the upper 100 feet or so of a glacier is brittle and is being carried along, as if on a conveyor belt, by the plastic, flowing ice below. The brittle ice becomes fractured as the flowing ice changes velocity, flows over cliffs, or makes sharp turns in the glacial channel (fig. 5). Ice at the center of a glacier moves faster

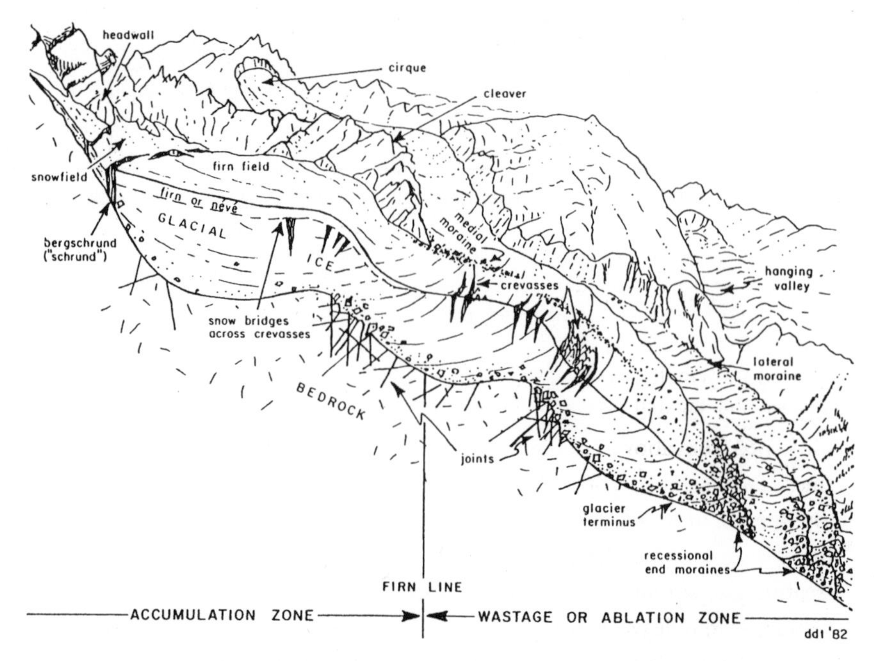

Figure 5. Features of a valley glacier. A cirque glacier is shorter but has many similar features. (Drawing by D. D. Trent; Trent, 1983.)

than ice at the sides, and this difference in speed also creates crevasses. Crevasses are open where brittle ice is pulled or stretched; they may later close when the ice moves into an area where compression is occurring.

Crevasses can be deadly, and no one should explore a glacier without being roped to a companion. The hope is that only one person at a time will fall into a crevasse. Despite appearances and war stories, crevasses are never bottomless. A mountaineer stumbling into one can reflect on the way down that the fall will not likely exceed 100 feet because, below about that depth, flow of ice prevents a crevasse from forming. Crevasses are rarely even 100 feet deep because often they are partially filled with snow. But a crevasse of any depth can be difficult or impossible to get out of without help from above, even if a person is strong and unhurt.

GLACIER ADVANCE AND RETREAT

Figure 5 is a cross section along the length of a valley glacier showing a zone of accumulation and a zone of *ablation,* or wastage. Ablation means loss of snow and ice by melting, evaporation, or any other means. The boundary between zones of accumulation and ablation is called the *firn line,* or *equilibrium line.* At the end of the summer melting season, snow remains in the accumulation zone while bare glacier ice is visible in the ablation zone (fig. 4). Snow in the accumulation zone becomes firn, then ice, which is added to the glacier. The ice flows downhill into the ablation zone, where it melts. The health and behavior of a glacier are determined by the balance, or lack of balance, between gain in the accumulation zone and loss in the ablation zone. The larger the area of the accumulation zone compared to the ablation zone, the healthier the glacier. A glacier in equilibrium has about two-thirds of its area in the accumulation zone and one-third in the ablation zone. Inspection of figure 4 shows that the Palisade Glacier is short of this ideal, as are most California glaciers in the present climate.

If over a period of many years there has been decreased snowfall (or increased snowmelt), the amount of snow added to the glacier will be reduced and the glacier will find itself with an ablation zone larger than can be replenished from above. Under such conditions the glacier is too long, causing the position of the terminus of the ice to move uphill, or backward; then we say the glacier is retreating. But even in a retreating glacier the ice never flows backward; it is only the location of the terminus that changes. A retreating glacier is likely to have a gentle slope at its terminus, usually concave upward, and to leave behind isolated masses of stagnant ice.

Conversely, more snow added to the accumulation zone causes a glacier to advance and reach a lower elevation before melting. Advancing glaciers commonly have steep convex-upward slopes at their terminus and, like a bulldozer, may push piles of sediment ahead and knock down trees.

The size and the behavior of glaciers are influenced by many factors, but ultimately these are determined by the balance between snowfall and snowmelt and they reflect variations in climate. When we study glacial advances and retreats, ancient or modern, we learn about climate changes past and present. Small temperate glaciers such as those in California are more sensitive to climatic change than are large glaciers, and observing them is a good way to monitor changing climate. Study of Ice Age glaciers helps increase our understanding of natural climate change and can perhaps help us distinguish human-induced change from natural fluctuation. Such changes are important far beyond the mountains where they are observed, because climate influences water supply, food production, energy production and consumption, and general economic well-being.

THE ICE AGE

During much of the last 1.5 million years, glaciers covered millions of square miles of continents that today are ice free, and large glaciers existed in mountain ranges that today have only small glaciers or none at all. This was the Ice Age. The idea of an Ice Age was first expressed in Europe in 1837 when Louis Agassiz described evidence, collected by others, that glaciers in the Alps had at some time in the past extended much farther down their valleys, perhaps beyond the mountains onto the plains. Agassiz generalized these observations by suggesting "a great ice period" when glaciers once covered much of Europe and Asia. The persons who had made the key observations had attracted little notice, but Louis Agassiz was a well-known and established scientist whose ideas could not be ignored, though many scientists of the time were skeptical. Charles Matsch presents a good summary of these interesting times in his chapter "The History of a Preposterous Idea" (1976).

In response to Agassiz, scientists and naturalists around the world put on boots and went into the field seeking evidence to support or refute the idea of an Ice Age. When the trail of an Ice Age glacier was first found in California, only twenty-six years had passed since Agassiz first expressed the idea of an Ice Age, and there was still powerful opposition to the idea. The influential English geologist Roderick I. Murchison insisted that glaciers could not erode rock, a view that was widely held among scientists everywhere, and one much studied and debated in California. Ignorance and uncertainty were widespread, even among believers in the new idea. There was much to be observed, learned, and unlearned, and many mistakes to be made and corrected, before the infant science of glaciers would mature. Much learning and sorting out would be done in the mountains of California, as will be related in later chapters of this book.

ICE AGES, PLURAL

When the idea of an Ice Age originated it was thought to be a unique event during the last million years of Earth's history. Subsequently, evidence was found of glaciations at other times in the more distant geologic past, and geologists began to speak of Ice Ages, including the one during the last million years and those that occurred hundreds of million of years ago.

Even though numerous Ice Ages are known to have occurred, people still speak as if there were only one. When we speak of the Ice Age we mean the one originally described, that of Agassiz. This most recent Ice Age is also called the Pleistocene Ice Age after the subdivision of geologic time during which most of the glaciation occurred (table 1). The Pleistocene Epoch began 1.6 million years ago and ended 10,000 years ago. For many years the names Pleistocene and Ice Age were used as synonyms, but such usage is misleading because Ice Age glaciers formed in some places before the beginning of the Pleistocene Epoch and many other things happened during the Pleistocene besides glaciation. Nonetheless, the name Pleistocene Ice Age is still commonly used. California has only a sketchy record of one possible ancient Ice Age, and it will not be considered in this book. This book is about the last Ice Age (the Pleistocene Ice Age) and events that have occurred since it ended 10,000 years ago.

TABLE 1. The last 65 million years of geologic time. We live in the Holocene Epoch of the Quaternary Period of the Cenozoic Era. The Ice Age began shortly before the Pleistocene Epoch, which began about 1.6 million years ago and officially ended at the close of the Pleistocene, 10,000 years ago. Modern glaciers in California formed in the Holocene Epoch during a time called the Little Ice Age, which began about 700 years ago.

ERA	PERIOD	EPOCH
Cenozoic	Quaternary	Holocene (Recent)
		Pleistocene
	Tertiary	Pliocene
		Miocene
		Oligocene
		Eocene
		Paleocene

MULTIPLE GLACIATION DURING THE ICE AGE

Glaciers advanced and retreated many times during the Ice Age, and it is these fluctuations driven by fickle climate that we have in mind when we speak of multiple glaciation. The times when glaciers became large are called *glacial advances, glacial episodes,* or *glacial stages,* and they have been named. Times when glaciers retreated or disappeared are called *interglaciations.* For example, the first glacial episode of the Ice Age known in California began about 1.5 million years ago and is called the McGee glaciation. There followed an interglaciation, then several other major glacial advances and interglaciations. Glaciers may or may not have completely disappeared from California during the interglaciations. The final significant Ice Age glacial advance reached a maximum about 20,000 years ago, after which glaciers quickly retreated. It is unlikely any glaciers were left in California 10,000 years ago when the Pleistocene Epoch ended.

IS THE ICE AGE OVER?

Yes. The Ice Age and the Pleistocene Epoch officially ended 10,000 years ago, at which time the Holocene (also called the Recent) Epoch began. But a more pertinent question to ask is . . .

WILL THERE BE ANOTHER MAJOR GLACIATION?

Most likely, yes. Though the Pleistocene Ice Age is, by definition, finished, there is no reason to believe that glaciations are extinct like the dinosaurs. There is good reason to believe that large glaciers will return at some future time.

Millennia after the end of the Pleistocene, glaciers formed again in California. Has the next major glaciation begun? Not necessarily. After a warm period in the early Holocene, world climate cooled, glaciers that had survived the warm spell grew larger, and new glaciers formed in California and elsewhere in temperate climates. This cooling, called the Little Ice Age, began about 700 years ago. The end of the Pleistocene Ice Age was the beginning of an interglaciation, a time of warmth; but climate fluctuates, and the Little Ice Age is believed to be a minor cooling superposed on a longer warm period. Indeed, the last 100 years have seen warmth return and glaciers recede around the world.

Civilization has developed during, and we are living in, a major interglaciation, one of perhaps a dozen that have occurred during the last million years or so. Each interglaciation has been followed by another major glacial advance, and there is no reason to believe that the present interglaciation will be any different. Emiliani cites evidence from deep-sea cores that "temperatures as high as those of today occurred for only about 10% of the time during the past half million years" (1972). Based on averages derived from study of the durations of warm and cold spells from the geologic record, Mörner presents informed speculation that the "Present Interglacial" started 12,700 years ago and will last another 18,800 years, at which time "the First Future Glacial Age is supposed to begin" (1972). It is unknown, of course, if this prediction will prove to be correct, but the prediction has more substance to it than a simple guess.

Another major glacial advance will be a big deal. Glacier ice covered as much as 32 percent of the land on Earth at various times in the Pleistocene. During the last 100,000 years, people not that different from you and me were forced to migrate because of advancing ice. However, Earth's population was small then. A future glacier advance of the magnitude of the one that peaked about 20,000 years ago will be of vastly greater impact on a crowded planet with complex social organization. One can expect that food production will decline dramatically, hundreds of millions of people will have to relocate, nations will become unstable, wars will be fought for habitable land, and civilization will be threatened. Some nations might benefit from the change of climate, but their good fortune will surely be offset by threat of war or conquest.

CAUSE OF ICE AGES

Why do Ice Ages occur? The subject is quite complex and no final answer is agreed to by all, but here is a short answer that seems to be on the right track. It appears that Ice Ages are of astronomical origin modified by certain geologic circumstances.

Astronomical factors determine the amount of solar heating received in the northern hemisphere, and this amount changes with time (Milankovitch, 1941). Three important aspects of Earth's relation to the Sun cause the variation: (1) change in the tilt of Earth's axis of rotation with respect to its plane of revolution (a 41,000-year cycle), (2) degree of ellipticity of Earth's orbit around the Sun (a 95,800-year cycle), and (3) precession of the equinoxes (the time of year the northern hemisphere is most tilted toward the Sun, a 21,700-year cycle). Each of these influences the amount of heating received by the northern hemisphere, and over time each interacts with the others to produce a complex variation in heating. Sometimes the three factors reinforce one another to produce exceptional cooling, while at other times they interact to produce strong warming. When they work against, and cancel, one another, in-between temperatures occur. The result is that the amount of solar heating received in the northern hemisphere varies with time in a complex but predictable manner.

We can calculate how solar heating has varied for the past tens of thousands of years. The best record of climate fluctuations during the Ice Age is contained in the sediment record of the ocean basins. This record has been compared with the variation in solar radiation calculated to have occurred as a result of the interacting astronomical cycles, and there is close agreement (Broeker, 1966; Hays, Imbrie, and Shackleton, 1976). This good correlation suggests a cause-and-effect relationship between astronomical cycles, climate change, and glacier behavior during the Ice Age.

But the correlation just described is not perfect, and it appears that something else is involved. Geologic events may work with or against the background astronomical pattern to produce glacial advances and retreats on the continents, and their role may be decisive at times. Some important geologic factors are volcanic eruptions, uplift of great mountain ranges that changes atmospheric circulation, changes in circulation of ocean water, and changing positions of continents caused by plate tectonics.

Here is an example of how astronomical cycles and geologic events might interact. Volcanic eruptions are known to cause cooling because they blast gases and particles of ash high into the atmosphere that can remain for years and cause a significant decrease in the amount of solar radiation reaching Earth's surface. Suppose that at some particular time the astronomical circumstance is appropriate for a small cooling, but insufficient to cause a glaciation. A large volcanic eruption occurring at that time might augment the astronomical cooling enough so that glaciers begin to form. Once formed, glaciers reflect much sunlight back to space and influence weather in such a way as to tend to perpetuate themselves. The result is a major glaciation that neither the astronomical situation nor the volcanic eruption alone could cause. An identical volcanic eruption at a time when the astronomical circumstance called for warming would not be capable of causing a glacial advance. According to this line of reasoning no single event, astronomical or geologic, will correlate with the glacier record perfectly. This rings true because research whose goal has been to find such a single cause has failed.

California, incidentally, has produced at least one volcanic eruption of a size possibly sufficient to alter world climate. This was the eruption 738,000 years ago of the Bishop Tuff from the Long Valley caldera east of the Sierra Nevada. This great eruption deposited volcanic ash over twenty-three states, as far east as Arkansas. Vast quantities of ash and gases must have been in the stratosphere for years thereafter.

Those who wish to learn more about this subject may read the following: Imbrie and Imbrie (1979) give an interesting overview of the cause of Ice Ages; and Bray (1979), Porter (1986), and Scuderi (1990) give information about how volcanic eruptions during the last few thousand years are closely related to glacier advances.

REFERENCES CITED

Bray, J. R. 1979. Surface albedo increase following massive Pleistocene explosive eruptions in western North America. *Quaternary Research* 12: 204–11.

Broeker, W. S. 1966. Absolute dating and the astronomical theory of glaciation. *Science* 151: 299–304.

Clark, Douglas H., Malcolm M. Clark, and Alan R. Gillespie. 1991. A late Pleistocene ice field in the Mokelumne drainage, north-central Sierra Nevada. *Geol. Soc. of Amer. Abstracts with Programs* 23, no. 2: 13.

Emiliani, Cesare. 1972. Quaternary hypsithermals. *Quaternary Research* 2: 270–73.

Hays, J. D., J. Imbrie, and N. J. Shackleton. 1976. Variations in the earth's orbit: Pacemaker of the Ice Ages. *Science* 194: 1121–32.

Imbrie, John, and Katherine Palmer Imbrie. 1979. *Ice ages: Solving the mystery.* Short Hills, N.J.: Enslow.

Kehrlein, Oliver. 1948. California's frozen water supply (glaciers). *Pacific Discovery* 1: 18–25.

Matsch, Charles L. 1976. *North America and the great Ice Age.* New York: McGraw-Hill.

Milankovitch, M. M. 1941. *Canon of insolation and the Ice-Age.* Special Publication 132 of the Royal Academy of Serbia, Belgrade. English translation by the Israel Program for Scientific Translations, published for the U.S. Dept. of Commerce and the National Science Foundation, Washington, D.C. (1969).

Mörner, Nils-Axel. 1972. When will the present interglacial end? *Quaternary Research* 2: 341–49.

Porter, Stephen C. 1986. Pattern and forcing of northern hemisphere glacier variations during the last millennium. *Quaternary Research* 26: 27–48.

Scuderi, L. A. 1990. Tree-ring evidence for climatically effective volcanic eruptions. *Quaternary Research* 34: 67–85.

Trent, D. D. 1983. The Palisade Glacier, Inyo County. *Calif. Geology* 36: 264–69.

Wahrhaftig, Clyde, and J. H. Birman. 1965. The Quaternary of the Pacific mountain system in California. In *The Quaternary of the United States,* edited by H. E. Wright and D. G. Frey, 299–340. Princeton, New Jersey: Princeton Univ. Press.

PART 2

ICE AGE GLACIERS

CHAPTER TWO

DISCOVERY AND GLACIAL HISTORY

DISCOVERY

Use of the words *glacier* and *California* in the same sentence began in the summer of 1863 when Josiah D. Whitney, William H. Brewer, and Charles F. Hoffman of the newly created Geological Survey of California visited Tuolumne Meadows in what is now Yosemite National Park.

> We have found so much of interest here, among the rest finding the traces of enormous *glaciers* here in earlier times, the first found on the Pacific slope, that we have been detained much longer than we expected.
>
> (Brewer, 1966, p. 409)

> [T]he whole region about the head of the Upper Tuolumne is one of the finest in the State for studying the traces of the ancient glacier system of the Sierra Nevada. . . . [A]ll of the phenomena of the moraines—lateral, medial, and terminal—are here displayed on the grandest scale.
>
> (Whitney, 1865, p. 428)

> A great glacier once formed far back in the mountains and passed down the valley, polishing and grooving the rocks for more than a thousand feet up on each side, rounding the granite hills into domes. It must have been as grand in its day as any that are now in Switzerland. But the climate has changed, and it has entirely passed away. (Brewer, 1966, p. 412)

A great glacier indeed! The Tuolumne Glacier formed in a highland ice field that began at the 13,000-foot peaks around Mount Lyell. The ice flowed along the Lyell Fork of the Tuolumne River, through Tuolumne Meadows (fig. 6), Muir Gorge, the Grand Canyon of the Tuolumne, Hetch Hetchy Valley, and

Figure 6. Tuolumne Meadows, looking east from Pothole Dome. The Tuolumne Glacier was about 2,000 feet thick here and flowed toward the viewer.

beyond, gathering tributary ice streams along the way, a distance of 60 miles to its terminus at an elevation of 2,000 feet. The Tuolumne Glacier was over 4,000 feet thick in places, and some ice overflowed through low passes into the Merced River Basin and Yosemite Valley.

Many visitors to Tuolumne Meadows climb Pothole Dome and Lembert Dome (fig. 7) over rock polished smooth by the Tuolumne Glacier, which covered these landmarks 20,000 years ago. Even those who stay close to their automobiles note the unusual shape of both domes. Each has a gentle streamlined slope where it faces up-valley, against the flow of the ice, but steep and rough sides facing down-valley where ice flowing away from the hill pulled, or plucked, blocks of rock loose from the mountain and carried them away. A hill, or any size mass of rock, having this distinctive shape created when it was overridden by ice is called a *roche moutonnée* (fig. 8). The distinctive shape of these glacial landforms is a consequence of the nature of rock, which is strong and durable under compression (when pushed) but weak under tensile stress (when pulled). Roches moutonnées of all sizes are abundant in the Sierra Nevada and some other California mountains, are easy to spot, and provide good evidence that glaciers once were present.

Moraines, mentioned in the quote from Whitney, tell us the size and extent of former glaciers. *Moraines* are piles, ridges, or irregular layers of intermixed sand, gravel, and boulders called *till* (fig. 9). Till forms from rock debris that either falls onto a glacier from adjacent cliffs or is scraped loose

Figure 7. Lembert Dome at Tuolumne Meadows is a large roche moutonnée. The Tuolumne Glacier flowed over the dome from right to left and removed perhaps half of the rock originally present.

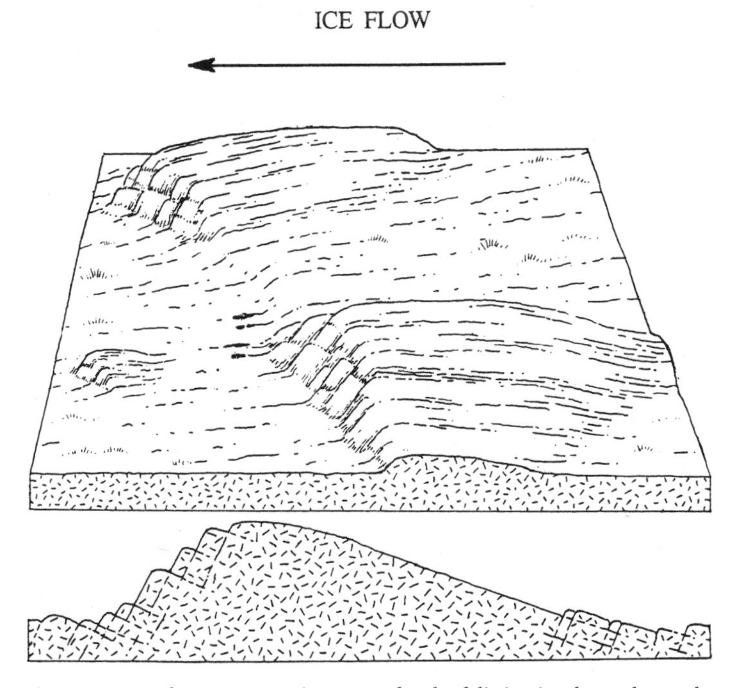

Figure 8. A roche moutonnée is a mass of rock of distinctive shape, the result of being overridden by a glacier. The smooth, gentle slope faced upstream and the steep, rough slope faced downstream. (Drawing by Susan Van Horn.)

Figure 9. A typical exposure of till east of the Sierra crest. Till is an unsorted, or poorly sorted, mixture of clay, sand, gravel, and boulders, deposited by a glacier.

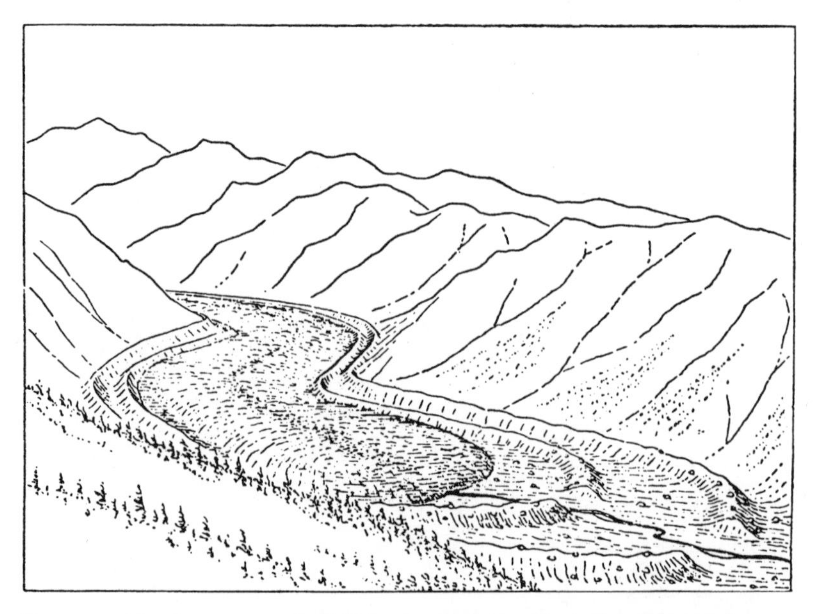

Figure 10. Moraines are masses of till that accumulate on, in, under, and around a glacier. Lateral moraines form on the sides; an end moraine extends around the end of a glacier connecting the laterals. A terminal moraine is an end moraine at the maximum advance of the ice. A retreating glacier may leave a series of end moraines called recessional moraines. (Drawing by U.S. Geological Survey; Matthes, 1930.)

by the moving ice. A glacier commonly deposits till along its sides and at its end, leaving a horseshoe-shaped moraine that is an outline of the former glacier. Moraines can endure long after the ice has melted. When a moraine is in the form of a ridge, the most common form, we can say that it is a "hill of till." Figure 10 shows the main types of moraine ridges. Till deposited as a thin irregular layer under the ice forms a *ground moraine.*

EXPLORATION

After finding evidence of Ice Age glaciers at Tuolumne Meadows in the summer of 1863, members of the Geological Survey of California explored throughout the Sierra Nevada and found evidence of former glaciers everywhere. The Geological Survey had been established in 1860 with Josiah D. Whitney as state geologist. William H. Brewer was in charge of work in the field; Whitney joined the field parties on occasion as other duties permitted. A young geologist named Clarence King soon joined the Survey and began a career that would include worldwide fame for his book *Mountaineering in the Sierra Nevada* and for being first director of the United States Geological Survey. When he joined the Whitney Survey, King was only twenty-one years old and he held a degree in chemistry from Yale, where he had acquired some knowledge of glacial geology from Louis Agassiz himself.

In July 1864, a California Geological Survey party of five, with Brewer in charge, explored the headwaters of the Kings River. Brewer wrote,

> We struck a ridge which is a gigantic moraine left by a former glacier, the largest I have ever seen or heard of. It is several miles long and a thousand feet high. (Brewer, 1966, p. 524)

The next day Brewer and Hoffman climbed a high peak (later named Mount Brewer) to get bearings and scout the Sierra crest:

> Such a landscape! A hundred peaks in sight over thirteen thousand feet—many very sharp—deep canyons, cliffs in every direction almost rivaling Yosemite, sharp ridges almost inaccessible to man, on which human foot has never trod—all combined to produce a view the sublimity of which is rarely equaled, one which few are privileged to behold. (Brewer, 1966, p. 524)

The landscape described is characteristic of that produced by extensive erosion of mountains by glaciers (figs. 11–13). Prior to glaciation the Sierra Nevada had a landscape typical of one shaped by rain and river erosion, with

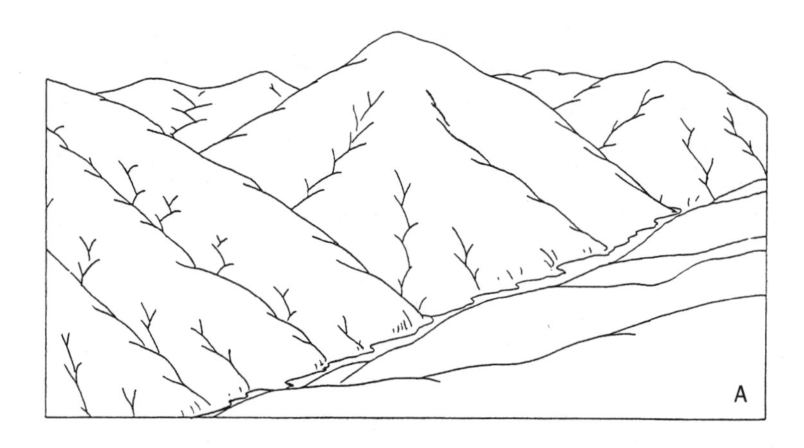

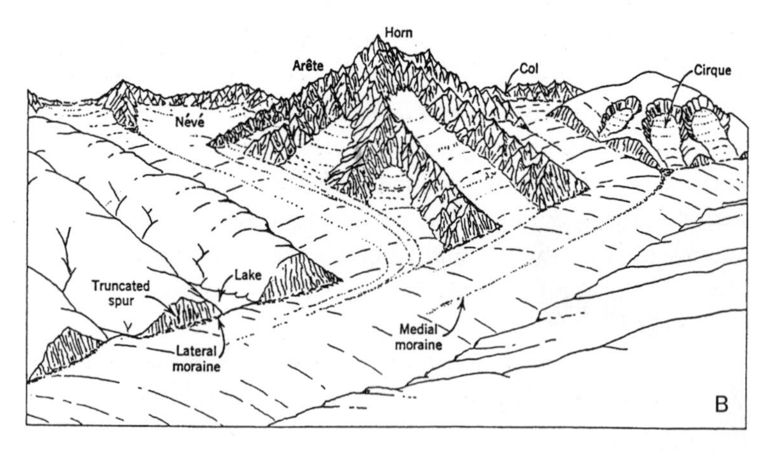

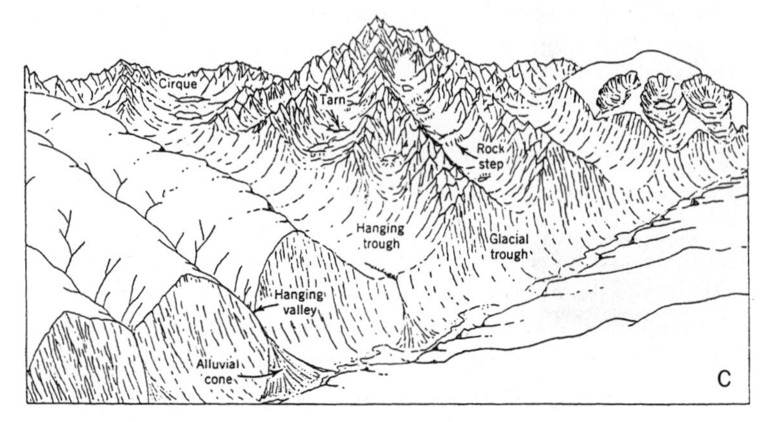

Figure 11. How glaciers change a landscape. A. Before glaciation. B. During glaciation. C. After glaciers melt. (Drawing and copyright by Arthur N. Strahler; Strahler, 1951.)

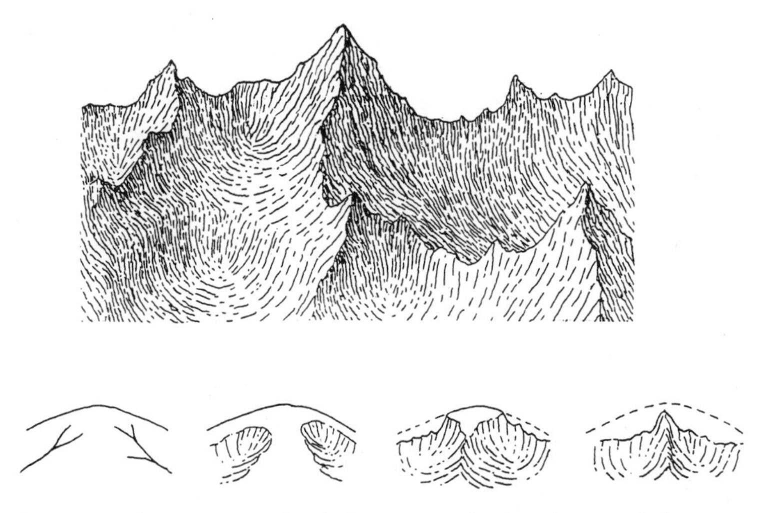

Figure 12. A horn is a pointed peak that remains after the enlargement of cirques largely consumes a preglacial mountain. (Drawing by Susan Van Horn.)

Figure 13. Horns at the head of Bishop Creek, east side of the Sierra Nevada. (Photo by George Hafer.)

rolling hills, gentle slopes, and few cliffs. Repeated glaciation produced the landscape Brewer described, with steep slopes, pointed peaks, and broad, open valleys. The sharp peaks are called *horns* and the narrow ridges between horns are called *arêtes.* A *col* is a low point on an arête, a pass that mountaineers sometimes cross. A *tarn* is a small lake formed by a glacier; some tarns lie in basins eroded by ice, others were formed when a glacier deposited a moraine that dammed a creek. A glacial lake in a cirque is a tarn, and it is also called a *cirque lake.*

Clarence King persuaded Brewer to let him and Dick Cotter try to climb the highest of the peaks seen from Mount Brewer, which all believed to be the highest in California. Thus began a great adventure, "a campaign for the top of California" described in *Mountaineering in the Sierra Nevada* in the chapter "The Ascent of Mount Tyndall." King noted many glacial features on the way, including "rude piles of erratic glacial blocks" and "the great Kings Cañon with its moraine-terraced walls." From the Kings-Kern Divide he looked south down the valley of the Kern River,

> with its smooth oval outline, the work of extinct glaciers, whose form and extent were evident from worn cliff-surface and rounded wall; snowfields, relics of the former névé, hung in white tapestries around its ancient birthplace; and, as far as we could see, the broad, corrugated valley, for a breadth of fully ten miles, shone with burnishings wherever its granite surface was not covered with lakelets or thickets of alpine vegetation. (King, 1872, p. 84)

This is a fine description of a glacial valley formed by a glacier widening, straightening, and eroding a river valley. The "oval outline" refers to the U-shaped profile so typical of glacial valleys (figs. 14 and 15). A névé (fig. 11) is the birthplace of a glacier, above the snow line, where old snow is converted into firn and glacial ice by recrystallization and compaction (*névé* also is used as a synonym for *firn*). *Burnishing* refers to rock polished by flowing ice (fig. 16). Glacial polish is produced on rock by abrasion by mud under high pressure at the base of a glacier.

Onward the men went, over rocks "so glacially polished and water-worn that it seemed impossible," eventually reaching easier going where "sheltered among roches moutonnées" there began to appear

> [l]akelets, small but innumerable, . . . held in glacial basins, the striae and grooves of that old dragon's track ornamenting their smooth bottoms. (King, 1872, p. 89)

Striae and grooves form under a glacier in the same way that polish does, but they are shaped by sand, gravel, and boulders in the ice instead of mud.

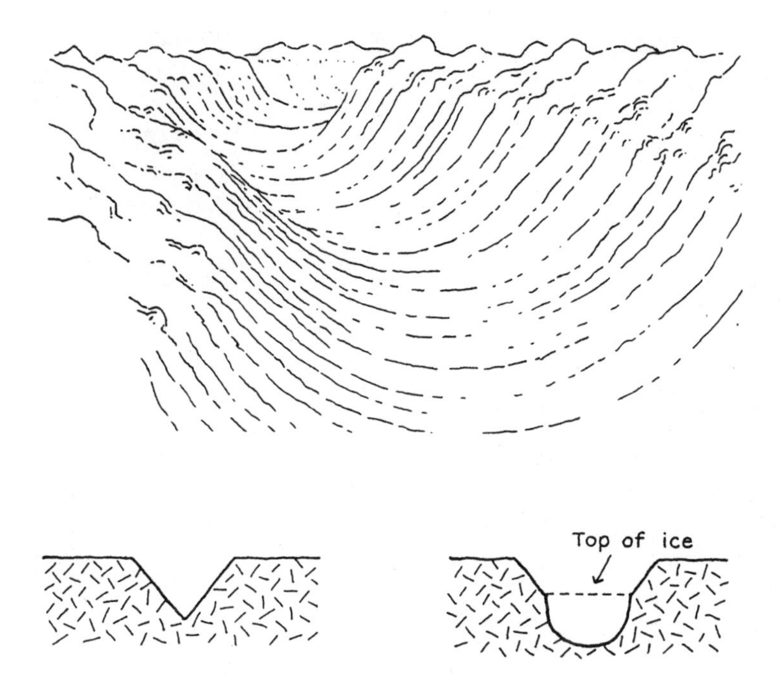

Figure 14. A U-shaped glacial valley is broader, more open, and straighter than a V-shaped river valley. The former top of the ice is marked by a trimline above which exposed rock is rough and below which exposed rock is smooth and rounded. (Drawing by Susan Van Horn.)

Figure 15. The U-shaped profile of the valley of Pine Creek, east side of the Sierra Nevada, north of Bishop.

Figure 16. Glacial polish and grooves at Tenaya Lake, Sierra Nevada. Polishing occurs where a glacier slides over underlying bedrock.

The landscape evoked in King a vivid image of the Ice Age, a feeling that many who have visited the high peaks above timberline will recognize:

> Something there is pathetic in the very emptiness of these old glacier valleys, these imperishable tracks of unseen engines. One's eye ranges up their broad, open channel to the shrunken white fields surrounding hollow amphitheaters which were once crowded with deep burdens of snow—the birthplace of rivers of ice now wholly melted; the dry, clear heavens overhead, blank of any promise of ever rebuilding them. I have never seen Nature when she seemed so little "Mother Nature" as in this place of rocks and snow, echoes and emptiness. It impresses me as the ruins of some bygone geological period, and no part of the present order, like a specimen of chaos which has defied the finishing hand of Time. (King, 1872, p. 98)

King and Cotter climbed their peak, but it was not the top of California; a higher peak, later named Mount Whitney, was seen from the summit.

The summers of 1863 and 1864 were golden years of High Sierra exploration. Studies by the California Geological Survey established that large glaciers once were widespread throughout the Sierra Nevada as well as the Cascade Mountains to the north. People began to wonder what the extent of former glaciation in California was.

THE NORTHERN ICE SHEET

Recognition that there once had been a great continental glacier in North America began in 1841 in Massachusetts. It was soon recognized that during the Ice Age an enormous glacier covered much of Canada and flowed south into the United States. Did the great northern ice sheet extend as far south as California? Were the California glaciers part of the great continental glacier? Joseph LeConte of the young University of California in Berkeley thought the southern margin of the northern ice sheet was in northern California, but north of the Sierra Nevada. John Muir, however, wrote:

> All California has been heavily glaciated, the broad plains and valleys so warm and fertile now, and the coast ranges and foothills. . . .
>
> Go where you may, throughout the length and breadth of the state, unmistakable evidence is everywhere presented of the former existence of an ice-sheet, thousands of feet in thickness, beneath whose heavy folds all the present landscapes have been molded; while on both flanks of the Sierra we find the fresher and more appreciable traces of the individual glaciers, or ice-rivers, into which that portion of the ice-sheet which covered the range was divided toward the close of the glacial period. (Muir, 1880, p. 550)

Josiah D. Whitney disagreed with LeConte and Muir:

> The explorations of the Geological Survey of California have demonstrated, however, that there is no true Northern Drift within the limits of this State. . . . There is nothing anywhere in California which indicates a general glacial epoch, during which ice covered the whole country and moved bodies of detritus over the surface independently of its present configuration, as is seen throughout the Northeastern States. (Whitney, 1882, p. 73)

"Northern Drift" refers to moraines, till, and other sediment deposited by the continental glacier. Today we know that two great ice sheets covered Canada and a large part of the northeastern and north-central United States during the Ice Age. The ice sheet did not extend nearly so far south in the western United States as it did east of the Rockies. Of the Pacific States, only northern Washington was under the ice sheet. Glaciers in southern Washington, Oregon, and California were not connected with the continental glacier but instead formed independently in the various mountain ranges. The mountains were never completely buried by ice as often occurs in ice sheets and ice caps. The higher peaks of the Sierra Nevada were ice free even when adjacent valleys were filled with thousands of feet of ice. We know this because rock that was beneath the ice looks different from that which was above. Ice-scoured rock is smooth and rounded, whereas the cliffs above the ice have an angular and rough appearance because of exposure to air with repeated freezing and thawing, avalanches, and rockfalls.

MULTIPLE GLACIATION

Israel C. Russell studied glaciation west of Mono Lake and found that

> [t]here were at least two periods of marked extension and many minor fluctuations. What may be termed an interglacial period is recorded in the sediments of the ancient lake [Mono Lake] by an accumulation of gravel separating two heavy deposits of lacustral sediments. This is the record of a time when evaporation exceeded supply, and it may reasonably be supposed to indicate a period of aridity. It appears from the study of the moraines that there were many fluctuations in the ice-streams which formed them . . .
>
> (Russell, 1889, p. 392)

Other investigators found evidence of two glacial episodes, including Hershey in the Klamath Mountains, and Knopf and Matthes in the Sierra Nevada (Hershey, 1903; Knopf, 1918; Matthes, 1924). Matthes pointed out that the older of the two glaciations was the more extensive, and that it would probably prove "upon further study to embody a multiple record" in itself (1925). He also suggested that the two glacial episodes were separated by a very long period of time because of conspicuous differences in preservation of old and young moraines, absence of glacial polish in areas covered only by the old ice, great depth of stream erosion between the two episodes, and residual rock forms created by weathering between the two glaciations.

Evidence was subsequently found of three glacial advances, then four. Blackwelder named four glaciations the McGee, Sherwin, Tahoe, and Tioga, in order from oldest to youngest (1931). These names are still used today. He named the episodes after localities where typical deposits were found, instead of numbering them, because he suspected that future work might add other episodes in between.

Before continuing with the story of multiple glaciation during the Ice Age, let us talk about how different glaciations are recognized and dated.

HOW GLACIAL DEPOSITS OF DIFFERENT AGES ARE IDENTIFIED

Blackwelder described various ways of distinguishing moraines of different ages. He described one technique by saying:

> Weathering of boulders in the till . . . was found most useful. . . . Since the weathering of such rocks as quartzite, basalt, slate, granite, and marble can

> not be readily compared, it was found best to concentrate attention on a single type of rock—the average granodiorite of the Sierra Nevada batholith. Fortunately this occurs abundantly in most of the tills in the region. Boulders of this rock were then counted at certain places and classified as being (a) almost unweathered, (b) notably decayed on the surface but still solid, (c) greatly weathered, cavernous, or rotten. These figures constitute a ratio known in the field as the G.W.R. or "granite weathering ratio." A ratio of 90–10–0 would be sure indication of the latest age; one of 30–60–10 would be typical of the Tahoe stage; while one of 0–30–70 would be afforded by only the older tills. (Blackwelder, 1931, p. 877)

A similar technique is still used by glacial geologists in California and elsewhere. Such a procedure differentiates older from younger moraines and allows determination of the order in which moraines were deposited; this sort of dating is called relative dating and is quite valuable even though the age of the deposits in years is unknown.

Another relative age-dating technique utilizes soil profiles on moraines. In an area of somewhat uniform climate and rock type, such as on the east slope of the Sierra Nevada, soil development is largely a function of time, and older moraines will have better-developed soils. Examples of the use of soils for dating moraines are found in Berry (1994), Birkeland, Berry, and Swanson (1991), and Birkeland and Burke (1988).

A clever way to determine the amount of weathering a moraine has endured is to measure the speed with which sound waves travel through large granitic boulders found on the moraine:

> A new quantitative, reproducible method for determining relative ages of unconsolidated Quaternary deposits containing granitic clasts has been developed and tested. The technique makes use of a microseismic timer to determine the compressional wave velocity (clast-sound velocity) in each clast of a group chosen from a single deposit. From these data a group mean velocity is determined that is proportional to the age of the deposit; the youngest deposits having the highest velocities. . . . [T]he clast-sound velocity method may be used to determine ages of deposits up to one million years old when calibrated with sufficient radiometric dates and may also be used as a tool for correlating undated deposits. (Crook, 1986, p. 281)

In practice, one pounds the boulder (the clast) with a hammer; a switch on the hammer starts a timer at the instant of impact. A detector on the other side of the boulder stops the timer as soon as the compressional wave (a sound wave, or P-wave) generated by the hammer impact is received, and displays the elapsed time in fractions of a millisecond. Measuring the distance between impact point and detector and using simple arithmetic gives velocity.

> The method is based on the theory that as clasts weather, physical and/or chemical processes cause microfractures to develop in the clast with an accompanying increase in rock compressibility. This causes a concurrent decrease in the P-wave velocity. (Crook, 1986, p. 281)

The clast-sound velocity (CSV) technique proved to be effective on various moraines in the Mono Lake Basin.

> In general, CSV was the most effective relative dating method used in this study. It consistently distinguished at least three relative age groups in every canyon but one, yet did not contradict independent evidence that certain moraines or parts of moraines should be grouped together.
> (Bursik and Gillespie, 1991, p. A224)

VOLCANIC ROCKS

Glacial deposits are usually difficult or impossible to date directly, but volcanic rocks can often be dated by techniques utilizing radioactive elements. The age in years of glacial events can often be approximated if till is interlayered with volcanic materials, as is the case at places on the east side of the Sierra Nevada (fig. 17).

Figure 17. The boulders are part of the Sherwin Till. The light-colored material overlying the boulders is volcanic ash and pumice of the Bishop Tuff, which has been dated at 738,000 years old. Therefore the Sherwin glaciation occurred more than 738,000 years ago. Highway 395 about a mile north of Sherwin Summit, east of the Sierra Nevada.

Volcanoes often erupt fine powdery ash high into the air, as many people witnessed in 1980 when Mount St. Helens in Washington blasted ash over Washington, Oregon, and several states downwind. Ash that is erupted high into the atmosphere can be carried by wind for long distances and deposited over thousands or tens of thousands of square miles. A layer of ash a few inches thick makes an excellent time marker, and ash from different eruptions can be identified using special techniques of study. This distinctive layer of ash of known age preserved under one moraine and on top of a second may reveal their relative ages and maximum or minimum age in years.

AGE DATING USING RADIOACTIVE CHEMICAL ELEMENTS

Age of moraines in years can be determined by measuring the abundance of certain radioactive elements. Two methods described here are carbon dating and cosmogenic isotope dating.

Carbon dating can be done on any material that was once part of a living organism, for example, bone, shell, wood, or leaves. Trees are often killed by glacier advances, and careful inspection of moraines may reveal fragments of wood mixed in with the rock debris. The wood can be carbon-dated to determine how many years ago the tree was killed, and so the age of the moraine is established.

Here is how carbon dating works. All living substances contain the element carbon, which occurs in two forms, carbon-12 and carbon-14; the numbers refer to the masses of the atoms. Carbon-14 is radioactive and is produced in the atmosphere by cosmic rays. While a plant or animal is living, the ratio of the amounts of carbon-14 to carbon-12 in the organism is constant; but upon death, the amount of radioactive carbon-14 begins to decrease because the organism no longer can build it into itself from the environment. With time the ratio of carbon-14 to carbon-12 decreases in a known manner. About 5,730 years after death, half of the original carbon-14 will be left; after another 5,730 years only a fourth remains, and so on, the amount of carbon-14 decreasing by half every 5,730 years. Carbon dating is good only for objects up to 70,000 years old.

Cosmogenic isotope dating is a new and potentially very valuable technique.

> Direct dating of glacial landforms is intrinsically difficult because they are constructed out of older rocks, and most dating techniques measure the formation age of minerals rather than the age of geomorphic redistribution. . . .

> Therefore we have developed and applied a different method for dating glacial moraines and other landforms, accumulation of cosmogenic ^{36}Cl. . . . Production rates of ^{36}Cl in the subsurface are small because of the low underground radiogenic neutron flux. . . . When buried rocks are exposed at the land surface (for example, because of glacial excavation) ^{36}Cl begins to accumulate because of exposure to cosmic-ray activation.
>
> (Phillips and others, 1990, p. 1529)

To test the method, samples were collected from the tops of the largest boulders on the crests of various moraines at Bloody Canyon in the Mono Lake basin. The ratio of ^{36}Cl to normal chlorine was determined by a mass spectrometer. The more of the distinctive isotope found in the boulder, the older the moraine.

> The accumulation of chlorine-36 indicates that episodes of glaciation occurred at about 21, 24, 65, 115, 145, and 200 ka (thousand years ago). . . . [T]he timing of the glaciations correlates well with peaks of global ice volume inferred from the marine oxygen isotope record. . . .
>
> (Phillips and others, 1990, p. 1529)

The technique can be used with isotopes besides those of chlorine. Clark, Bierman, and Gillespie measured amounts of ^{10}Be and ^{26}Al from boulders and from glacially polished bedrock (1995). If the cosmogenic isotope technique proves to be consistent and accurate it will be a powerful tool in establishing glacial history.

MULTIPLE GLACIATION OF THE SIERRA NEVADA

Now back to multiple glaciation during the Ice Age. Blackwelder names and describes four glacial episodes and includes maps and photographs of glacial features characteristic of each (1931). Some excerpts from his descriptions follow:

Tioga Glaciation

> The glacial features that were made by the ice tongues of the Tioga epoch are even now almost as fresh and unaltered as at the time of their formation. . . . Acres of polished and grooved rock are a familiar sight. . . . The lateral moraines generally stand out as bold embankments. . . . The terminal moraines are still complete, except for V-shaped notches through which the main streams tumble down to the plains beyond. (Blackwelder, 1931, p. 882)

Tahoe Glaciation

The most conspicuous moraines in the Sierra Nevada are those of the Tahoe age. . . . The primary glacial features of the Tahoe epoch are still fairly well preserved and are easily recognizable, although they have been notably marred and obscured by the processes of weathering, stream erosion, and deposition. . . . In the bottom of the glacial troughs old roches moutonnées are easily recognizable by their rounded forms, but when they are examined for glacial polish and striations it is found that such surfaces have almost entirely disappeared. . . . Boulders are much less common on the Tahoe than on the younger moraines. . . . This may be due partly to the disappearance of boulders by decay, but in the writer's opinion is probably due to the accumulation of wind-blown dust, concealing most of the smaller or partly embedded boulders. (Blackwelder, 1931, p. 884)

Sherwin Glaciation

The Sherwin stage is now clearly represented by little except large bodies of unmistakable till. Upon these patches of glacial drift the boulders of igneous rock are all much weathered, most of them notably exfoliated, and many quite rotten. . . . Decay has penetrated deeply into the till. Boulders three feet thick have been readily cut through by the jaws of the steam shovels. . . . It would be difficult to recognize the glacial origin of the Sherwin till if it were not for the excellent exposures in fresh road excavations. There the wholly unstratified nature of the deposits is clearly revealed and one may readily find many stones scratched and facetted in the characteristic glacial manner. (Blackwelder, 1931, p. 896)

McGee Glaciation

The earliest of the four stages now recognized in the Sierra Nevada is represented only by isolated bodies of boulder earth, the original relations of which are now problematic. On the high ridge west of McGee Peak several thick patches of a deposit strongly resembling till and consisting largely of granite débris rest upon a foundation of Paleozoic slate and marble. . . . Since the material consists largely of granitic rock and yet rests upon metamorphic sedimentaries it must be a transported deposit. . . . It is also significant that these remnants now lie on the tops of divides, in situations where neither mudflow nor glacier could possibly emplace them, with topographic conditions as they are now. . . . [I]t is concluded that the deposits on Mount McGee antedate the present eastern front of the range with its deep canyons.
(Blackwelder, 1931, p. 902)

Blackwelder describes other localities where (presumed) McGee Till is found and cites reasons for believing the deposits are indeed glacial rather than of mudflow or landslide origin.

Since Blackwelder's four glaciations were named and accepted, other names have been proposed for what were believed to be other glaciations in the Sierra Nevada. Even more names have been given to glacial deposits in the Klamath Mountains and other glaciated areas of California. However, glacial names from the east-central Sierra Nevada have become the California standard, and researchers working elsewhere in California usually try to correlate the local glacial episodes with them. Only the Sierra Nevada names will be reviewed here.

At the same time that Blackwelder was working east of the Sierra crest, François Matthes was studying glacial deposits on the west slope in the Yosemite region. He recognized three glacial episodes, which he named the Wisconsin, El Portal, and Glacier Point, but these names have fallen into disuse, not because they are invalid, but because they are not needed. The glacial record on the east side of the Sierra Nevada is better preserved and easier to study than that on the west slope, and that is where the greatest interest is.

By the early 1970s the number of Ice Age glaciations recognized in the Sierra Nevada reached a maximum, as shown in table 2.

TABLE 2. Ice Age glaciations of the Sierra Nevada as understood in the early 1970s. Some of these are no longer regarded as valid. See table 3 for comparison.

Hilgard	about 10,000 years ago
Tioga	about 20,000 years ago
Tenaya	about 27,000 years ago
Tahoe	about 50,000 years ago
Mono Basin	about 130,000 years ago
Casa Diablo	about 400,000 years ago
Sherwin	about 750,000 years ago
McGee	about 2,500,000 years ago
Deadman Pass	about 3,000,000 years ago

Subsequent studies, aided by new techniques and age dates, have caused a number of changes to be made. The Deadman Pass Till, once thought to be one of the oldest in the world, is no longer regarded as a glacial deposit (Huber, 1981; Bailey, Huber, and Curry, 1990). Huber suggested the McGee is not as old as was once thought (1981). The validity of the Tenaya, Mono Basin, and Casa Diablo as separate glaciations has been questioned by Burke and Birkeland (1979). Clark and Clark present evidence that the Hilgard is part of the Tioga and not a separate glacial episode (1995). Clark, Heine, and Gillespie propose that a new glaciation, the Recess Peak, be added as the "last regional late-Pleistocene advance in the Sierra Nevada" (1996). The Recess Peak glaciation had previously been regarded as an advance that occurred after the end of the Pleistocene (see chapter 6).

Accepting the proposed revisions as valid, the current list of Ice Age glaciations in the Sierra Nevada, slim and trim, is shown in table 3. What happened to the Hilgard, Tenaya, Mono Basin, and Casa Diablo glaciations? Have they vanished into the geologic equivalent of a black hole? No. The fewer number of names does not mean that fewer glacier advances occurred. The moraines named Hilgard, Tenaya, Mono Basin, and Casa Diablo are still there and represent real glaciers; it is just that now they are regarded as part of the more important, longer-duration, expanded Tioga and Tahoe glacial episodes. The difference is in our minds, not in the mountains. The changes may appear minor, but to specialists the distinctions are important in that they clarify understanding and influence the direction of future research.

TABLE 3. Ice Age glaciations of the Sierra Nevada, 1996 minimalist version. Each glaciation other than the Recess Peak lasted a long time and included multiple advances and retreats. The ages in years are representative but do not reflect details of the episode or resolve conflicting data.

Recess Peak	about 14–15,000 years ago
Tioga	about 19–26,000 years ago
Tahoe	about 70–150,000 years ago
Sherwin	about 1,000,000 years ago
McGee	about 1,500,000 years ago

SUMMARY OF THE ICE AGE HISTORY OF CALIFORNIA

About 1.5 million years ago California climate became cool enough for glaciers to form in the Sierra Nevada (the McGee glaciation) and the Cascade Mountains. Probably there were a number of advances and retreats of the ice during this overall glacial time. The climate subsequently warmed, and glaciers retreated and may or may not have totally disappeared during the interglaciation. Glacial climate prevailed a second time (the Sherwin glaciation) about a million years ago. Like the McGee, there probably were numerous advances and retreats. The Sherwin glaciation appears to have been the most extensive in California history, and the one with the largest glaciers, and it probably affected geologic provinces other than the Sierra Nevada and the Cascade Mountains. Again an interglaciation followed, and in the next several hundred thousand years two more major glacial episodes (Tahoe and Tioga) occurred. Evidence of these last two glaciations is found in all provinces of California in which glaciation has been recognized (fig. 1). Multiple advances and retreats of ice during each of these glacial episodes produced a complex array of moraines on the east side of the Sierra Nevada that is still being unscrambled and dated. Tioga glaciers were smaller than Tahoe glaciers, which were smaller than Sherwin glaciers. Tioga glaciation reached its last maximum between 20,000 and 18,000 years ago, after which rapid retreat began. A final ice advance about 14,000 years ago (the Recess Peak glaciation) was the last Ice Age glaciation, with glaciers much smaller and much less extensive than Tioga glaciers. In all probability no glaciers were left anywhere in California when the Pleistocene Epoch, and the Ice Age, officially ended 10,000 years ago.

But the end of the Ice Age was not the end of the glacier story. Glaciers would return to California, as will be discussed in chapter 6.

REFERENCES CITED

Bailey, Roy A., N. King Huber, and Robert R. Curry. 1990. The diamicton at Deadman Pass, central Sierra Nevada, California: A residual lag and colluvial deposit, not a 3 Ma glacial till. *Geol. Soc. of Amer. Bull.* 102: 1165–73.

Berry, Margaret E. 1994. Soil-geomorphic analysis of late-Pleistocene glacial sequences in the McGee, Pine, and Bishop Creek drainages, east-central Sierra Nevada, California. *Quaternary Research* 41: 160–75.

Birkeland, P. W., and R. M. Burke. 1988. Soil catena chronosequences on eastern Sierra Nevada moraines, California, U.S.A. *Arctic and Alpine Research* 20: 473–84.

Birkeland, Peter W., Margaret E. Berry, and David K. Swanson. 1991. Use of soil catena field data for estimating relative ages of moraines. *Geology* 19: 281–83.

Blackwelder, Eliot. 1931. Pleistocene glaciation in the Sierra Nevada and Basin Ranges. *Geol. Soc. of Amer. Bull.* 42: 865–922.

Brewer, William H. 1966. *Up and down California in 1860–1864.* Edited by F. P. Farquhar. 3rd ed. Berkeley: Univ. of Calif. Press.

Burke, R. M., and Peter W. Birkeland. 1979. Reevaluation of multiparameter relative dating techniques and their application to the glacial sequence along the eastern escarpment of the Sierra Nevada, California. *Quaternary Research* 11: 21–51.

Bursik, Marcus, and Alan R. Gillespie. 1991. Relative ages of late-Pleistocene moraines in Mono Basin. *Geol. Soc. of Amer. Abstracts with Programs* 23, no. 5: A224.

Clark, Douglas H., Paul R. Bierman, and Alan R. Gillespie. 1995. New cosmogenic ^{10}Be and ^{26}Al measurements of glaciated surfaces, Sierra Nevada, California—

they're precise, but are they accurate? *Geol. Soc. of Amer. Abstracts with Programs* 27: A170.

Clark, Douglas H., and Malcolm M. Clark. 1995. New evidence of late-Wisconsin deglaciation in the Sierra Nevada, California, refutes the Hilgard glaciation. *Geol. Soc. of Amer. Abstracts with Programs* 27, no. 5: 10.

Clark, D. H., J. T. Heine, and A. R. Gillespie. 1996. Glacial and climatic complexities in the American Cordillera during the Younger Dryas period. *Geol. Soc. of Amer. Abstracts with Programs* 28, no. 7: 233.

Crook, Richard, Jr. 1986. Relative dating of Quaternary deposits based on P-wave velocities in weathered granitic clasts. *Quaternary Research* 25: 281–92.

Hershey, Oscar H. 1903. Some evidence of two glacial stages in the Klamath Mountains in California. *American Geologist* 31: 139–56.

Huber, N. King. 1981. *Amount and timing of late Cenozoic uplift and tilt of the central Sierra Nevada, California—evidence from the upper San Joaquin River Basin.* U.S. Geol. Survey Prof. Paper 1197.

King, Clarence. 1872. *Mountaineering in the Sierra Nevada.* Boston: James R. Osgood and Co.

Knopf, Adolph. 1918. *A geologic reconnaissance of the Inyo Range and the eastern slope of the southern Sierra Nevada, California.* U.S. Geol. Survey Prof. Paper 110.

Matthes, F. E. 1924. Evidence of two glacial stages in the Sierra Nevada. *Geol. Soc. of Amer. Bull.* 35: 69–70.

———. 1925. Evidences of recurrent glaciation in the Sierra Nevada of California. *Science,* n.s., 61: 550–51.

———. 1930. *Geologic history of the Yosemite Valley.* U.S. Geol. Survey Prof. Paper 160.

Muir, John. 1880. Ancient glaciers of the Sierra, California. *Californian* 2: 550–57.

Phillips, F. M., M. G. Zreda, S. S. Smith, D. Elmore, P. W. Kubik, and P. Sharma. 1990. Cosmogenic chlorine-36 chronology for glacial deposits at Bloody Canyon, eastern Sierra Nevada. *Science* 248: 1529–32.

Russell, Israel C. 1889. *Quaternary history of Mono Valley, California.* U.S. Geol. Survey 8th Annual Report, 261–394.

Strahler, Arthur N. 1951. *Physical geography.* New York: John Wiley and Sons.

Whitney, J. D. 1865. The High Sierra. Chap. 10 in *Report of progress, and synopsis of field work from 1860 to 1864.* Vol. 1 of *Geology.* Philadelphia: Geol. Survey of California.

———. 1882. The climatic changes of later geological times. *Contributions to American Geology.* Vol. 2. Cambridge, Mass.: Museum of Comparative Zoology.

CHAPTER THREE

ICE AGE GLACIERS OF THE SIERRA NEVADA

THE SIERRA NEVADA DURING THE ICE AGE

Let us go back in time to the Ice Age and take a tour of the Sierra Nevada from north to south, with François Matthes as our guide:

> [I]n the vicinity of Donner Pass, where the main peaks rise to altitudes of 8,000 and 9,000 feet, the glaciers of the latest stage attained lengths of ten to fifteen miles and coalesced to form a continuous ice mantle two hundred fifty square miles in extent. During the earlier stages the glaciers must have been larger. Donner Pass itself, 7,000 feet in altitude, was during each glacial stage completely submerged beneath a great ice sea, and long glaciers descended from it both westward and eastward.
>
> . . . [F]ifty miles south, where the main divide west of Lake Tahoe bears peaks 9,000 to 10,000 feet high, the later glaciers on the western slope were twenty miles long and the earlier glaciers were five to ten miles longer. East of the divide a row of cascading ice streams plunged into the basin of Lake Tahoe. And in the hundred mile stretch from Lake Tahoe to Yosemite National Park, in which the summit peaks rise progressively to altitudes of 11,000, 12,000, and 13,000 feet and the range attains a climax of ruggedness, there developed also a grand climax of glaciation. . . . [O]nly a few of the highest peaks projected as small dark "islands" of rock, *nunataks.* Along the axis of the range this . . . [highland ice field] . . . extended for eighty miles; in breadth it averaged forty miles; and from its margins long, tapering ice streams reached down the canyons in both the western slope and the eastern escarpment.
>
> Among those ice streams were the longest in the entire Sierra Nevada. In the canyons of the forks of the Stanislaus River they attained lengths of thirty miles during the latest glacial stage and of forty to forty-five miles during the earlier ages. Longest of all was the Tuolumne Glacier, which each time com-

pletely overwhelmed Hetch Hetchy Valley. During the latest glaciation it reached a length of forty-six miles, and during the earlier stages a maximum length of sixty miles. . . .

The Yosemite Glacier, the next ice stream to the south of the great Tuolumne Glacier, was by far the smallest trunk glacier in the central Sierra Nevada, for it headed in a separate and relatively small basin and received only moderate reinforcements by overflow from the neighboring ice cap. During the latest stage the Yosemite Glacier attained a length of only twenty-four miles and terminated in Yosemite Valley, just below Bridalveil Falls; but during the earlier ice ages it continued down the Merced Canyon a short distance below El Portal, thus reaching a length of thirty-six or thirty-seven miles.

Southeast of the Yosemite region, in the broad drainage basin of the San Joaquin River during the earlier ice ages lay another vast ice mass. It measured fifty miles in length along the axis of the range and thirty to thirty-five miles in breadth. Taken together with the adjoining ice masses in the basins of Dinkey Creek and the North Fork of the Kings River, it formed a . . . [highland ice field] . . . 1,500 square miles in extent. . . . It consisted of a large number of confluent glaciers that had descended from the surrounding peaks and crests, filled the canyons to overflowing, and spread over the intermediate uplands.

During the latest glaciation there was not enough ice to spread over so large a territory, and most of the glaciers lay confined in canyons. There were two great ice streams: the Middle Fork Glacier, which headed on Banner Peak and Mount Ritter, and the South Fork Glacier, which originated in Evolution Basin. The former measured thirty-three miles in length; the latter, forty-three miles. . . .

Entirely different was the situation in the drainage basin of the Kings River. Because of the tremendous depth of the Middle Fork and the South Fork canyons—5,000 to 7,000 feet—the trunk glaciers remained wholly separated and there could be no . . . [highland ice field] . . . spreading broadly across the Monarch Divide. But each of the two main ice streams headed in high-level valleys paralleling the crests of the range and filled with a continuous ice field twenty-four miles long; both ice streams received rows of cascading tributaries along their course.

During the latest glaciation the Middle Fork Glacier, which passed through Tehipite Valley, attained a length of twenty-eight miles, and the South Fork Glacier, reinforced by the ice that came down Bubbs Creek Canyon, attained a length of thirty miles. Yet these two trunk glaciers fell short . . . of reaching the junction of the canyons and remained wholly separate. . . .

Striking indeed must have been the appearance of the Kaweah Basin during glacial times, for each of its numerous converging canyons was the pathway of a turbulent, cascading ice stream. A dozen such ice streams, five to seven miles long, descended from the Great Western Divide, yet the largest single ice mass lay in the now clean-swept basin of exfoliating granite that extends from Tokopah Valley to the tableland at the head of the Marble Fork. This ice mass covered an area of about fifteen square miles and sent forth a trunk glacier ten miles long. These figures are for the later ice stream; the extent of the earlier ones is unknown and may never be known exactly, as the older moraines are concealed by dense chaparral on the lower slopes of the canyons.

The southernmost trunk glacier in the Sierra Nevada lay in the Kern Canyon. As that canyon extends in a nearly straight line through the middle of the Upper Kern Basin, and the tributary canyons branch from it like the ribs in an oak leaf from the main rib, the glacier system had the same pattern. During the latest glaciation the tributary ice streams coming down from the Great Western Divide on the west and the main Sierra crest on the east lay confined in the side canyons; but during the earlier glaciations there was more ice than the canyons could hold and it spread over the benchlands on either side of the main canyon to a total breadth of four to six miles, thus producing a . . . [highland ice field] . . . about thirty square miles in extent—a truly remarkable fact, considering that the entire expanse sloped southward and lay exposed to the rays of the midday sun.

. . . [D]uring the latest ice age the Kern Glacier attained a length of twenty-five miles, and during the preceding ice ages a length of about thirty-two miles. Hockett Peak marks approximately the southernmost limit of glaciation in the Sierra Nevada. Farther south the range was too low to bear glaciers. Its great glacier system, three hundred miles long, thus ended abruptly just beyond the southern limit of the High Sierra.

(Matthes, 1950a, p. 60)

SELECTED LOCALITIES

The rest of this chapter highlights places in the Sierra Nevada of special interest for one reason or another. Yosemite Valley is given special consideration in chapter 4.

TENAYA LAKE

Tenaya Lake, on the Tioga Road in Yosemite National Park, is a fine place to observe glacial landforms:

The traces of the existence of an immense flow of ice down the slopes of the valley . . . are exceedingly conspicuous. The ridges on both sides of the lake are worn and grooved by glaciers nearly up to their summits, and travelling over the pass, from the valley of the Tenaya into that of the Tuolumne, became very difficult for the animals, so highly polished and slippery were the broad areas of granite over which they were obliged to cautiously pick their way. The granite here has a very coarse texture, the crystals of feldspar being

Figure 18. Glacial polish and striations at Tenaya Lake. A light-colored aplite dike (lower left to upper right) in the granitic rock has been eroded flush with the granitic rock. Compare with figures 72 and 78.

> often three or four inches long. Above the sphere of glacial action, these crystals project from the weathered surface of the rock; but, below that, they are planed down to a level with it, and most beautifully polished.
>
> (Whitney, 1865, p. 425)

Glaciers polish rock over which they flow by the great pressure they exert and the presence of mud mixed in the ice. Tenaya Lake was at one time under as much as 2,460 feet of glacial ice that produced a pressure of about 74 tons per square foot and extensive glacial polish (fig. 18). Coarser sediment in the ice, such as sand and gravel, makes scratches or grooves rather than polish.

In 1870 and 1872 Professor Joseph LeConte led groups of students and graduates of the University of California, "The University Excursion Party," into Yosemite. LeConte also was impressed by the polished rock around Lake Tenaya:

> It is wonderful that in granite so decomposable these old glacial surfaces should remain as fresh as the day they were left by the glacier. But if ever the polished surface scales off, then the disintegration proceeds as usual. The destruction of these surfaces by scaling is in fact continually going on. Whitney thinks the polished surface is hardened by pressure of the glacier. I cannot think so. The smoothing, I think, prevents the retention of water, and thus prevents the rotting. . . . [R]otting of rock is hastened by roughness, and

> still more by commencing to rot, and retarded or prevented by grinding down to the *sound* rock and then polishing. (LeConte, 1870, p. 74)

LeConte was right about the exclusion of water aiding preservation of polished rock.

John Muir described glacial polish best:

> The most striking and attractive of the glacial phenomena in the upper Yosemite region are the polished glacier pavements, because they are so beautiful, and their beauty is of so rare a kind, so unlike any portion of the loose, deeply weathered lowlands where people make homes and earn their bread. . . . They are found in most perfect condition in the subalpine region, at an elevation of from eight thousand to nine thousand feet. Some are miles in extent, only slightly interrupted by spots that have given way to the weather, while the best preserved portions reflect the sunbeams like calm water or glass, and shine as if polished afresh every day . . .
>
> . . . The Indian name of Lake Tenaya is "Pyweak"—the lake of shining rocks. One of the Yosemite tribe, Indian Tom, came to me and asked if I could tell him what had made the Tenaya rocks so smooth. (Muir, 1912, p. 133)

When the Tioga Road was widened at Tenaya Lake during the 1950s large areas of glacially polished granite were destroyed, a circumstance that prompted the famous Sierra photographer Ansel Adams to submit his resignation as a director of the Sierra Club because the club did not oppose construction vigorously enough. Fortunately, considerable areas of polish remain.

TUOLUMNE MEADOWS

LeConte, like the Whitney Survey party before, saw abundant evidence of glaciation in Tuolumne Meadows (along with the presence of twelve to fifteen thousand sheep!), which inspired him to give a lecture on glaciers, profitably read even today (LeConte, 1870). In Bloody Canyon east of Tuolumne Meadows he found a marvelous natural laboratory for studying glacial lakes and described how glaciers form closed basins both by excavating bedrock and damming drainage by depositing moraines. In-filling of these lakes by waterborne sediment causes lakes to "graduate insensibly into marshes and meadows," a process that has formed meadows throughout the Sierra Nevada and that, unless the glaciers return or people intervene, is the fate of all Sierra lakes.

Two maps published by the U.S. Geological Survey show glacial features of Tuolumne Meadows. A geologic map shows the distribution of glacial deposits and the crests of moraines (Bateman, Kistler, Peck, and Busacca, 1983).

Another map shows the area, and all of Yosemite National Park, as it appeared about 20,000 years ago during the Tioga glaciation (Alpha, Wahrhaftig, and Huber, 1987). Tuolumne Meadows was covered by a highland ice field, and only the higher peaks, such as Mount Conness, Kuna Crest, and the Cathedral Range, were visible. The map shows overflow of ice into the basin of Tenaya Lake and from there into Yosemite Valley.

Malcolm Clark studied moraines at Tuolumne Meadows and arrived at an interesting conclusion:

> Tioga (latest Pleistocene) glaciers of the Sierra Nevada were destroyed rapidly by stagnation and disintegration while they were still quite extensive, rather than by gradual retreat of termini of actively moving streams of ice.
>
> In most drainages on both sides of the Sierran crest, distinct recessional moraines show gradual retreat of Tioga glaciers (with possible minor readvances) to 60–80 percent of maximum length. Between these recessional moraines and the position of either neoglacial moraines . . . or the Tioga cirques themselves, the only glacial deposits are scattered erratic boulders and very minor patches of till, outwash, or stratified drift.
>
> Some places, such as Tuolumne Meadows, contain many low, adjacent moraines alined in the direction of ice flow and located at about 30 percent of maximum Tioga length. Their position, orientation, shape, and small downglacier gradient indicate that these are medial moraines deposited by the ablation of large stagnant glaciers whose termini were far downvalley.
>
> . . . Stagnation and disintegration of these extensive Sierran glaciers indicate a major and very abrupt (perhaps within decades or centuries) change in climate in the Sierra Nevada near the end of Pleistocene time.
>
> (Clark, 1976, p. 361)

MONO LAKE AREA

> Mono Basin, a closed tectonic depression at the base of the steep Sierran escarpment east of Yosemite National Park, held a succession of deep lakes, collectively referred to as Lake Russell, during late Pleistocene time. Strand lines circle the basin below 7,180 feet, the altitude of the overflow channel into Adobe Valley to the south and ultimately into the Owens River. Mono Lake presently occupies the floor of the basin below 6,400 feet.
>
> (Lajoie, 1968, Abstract)

> The Mono Basin of eastern California provides an ideal laboratory in which to study the interaction of volcanic and tectonic processes. The late Quaternary record of volcanic activity and range-front faulting is relatively complete. Range-front faults of the Sierra Nevada offset dateable late Pleistocene glacial moraines, thus affording the opportunity to estimate range-front slip rates.
>
> (Bursik, 1988, Abstract)

Figure 19. A cirque is an amphitheater eroded into a mountainside at the head of a mountain glacier, where ice flowing away from the mountain pulls away blocks of jointed rock. (Drawing by Susan Van Horn.)

Here is a Quaternary geologist's dream. Glacial deposits, a pluvial lake, volcanoes, and faults, all forming and active at the same time. Because of this geologic wealth, the Mono Lake area is possibly the most studied part of the Sierra Nevada. It includes the Sierra crest in Yosemite National Park, the eastern slope of the range, and a major lake that was much larger during the Ice Age (a pluvial lake). Moraines are better preserved and displayed on the east slope of the Sierra Nevada than on the west slope, and the Mono Lake area has an especially good assemblage in several tributary valleys. Access is easy by Highway 395 and the many side roads that extend into the mountains to the west.

Quaternary History of Mono Valley, California is an 1889 publication by Israel C. Russell of such interest and value that it is still in print. Russell pre-

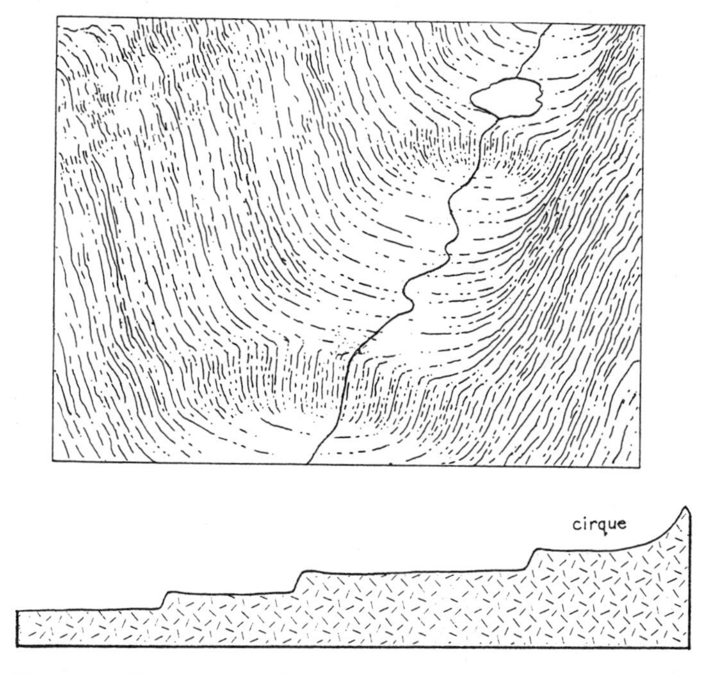

Figure 20. Glacial steps are cliffs in a glacial valley from which jointed rock has been plucked and carried away downstream. When a glacier is present there is an icefall. After the glacier melts there is usually a waterfall. (Drawing by Susan Van Horn.)

sents as introduction a description of a trip from the lowlands east of Mono Lake to the summit of Mount Dana on the Sierra crest and information about Mono Lake, before turning his attention to glacial features along the Sierra crest. His fine map shows the ancient glaciers west of Mono Lake, and detailed descriptions are given of those that occupied Lundy, Lee Vining, Gibbs, Bloody, and Parker canyons.

Russell described how glaciers create cirques, glacial steps, and rock-basin lakes. A cirque (fig. 19) is a three-sided bowl in a mountainside where a glacier began. The walls of a cirque are steep, perhaps vertical, and a lake is usually present, often touching the steep headwall. Glacial steps (fig. 20) are cliffs facing downstream in a glacial valley, with a gently sloping valley floor both above and below the cliff.

Putnam studied moraines and old shorelines at Mono Lake, which during the Ice Age was much larger and several hundred feet deeper than at present (1950). To honor the fine work Israel C. Russell did in the Mono Basin, Putnam named the large Ice Age predecessor of Mono Lake, Lake Russell.

Detailed maps by Putnam show moraines at Lee Vining Canyon, Grant Lake, and June Lake, and shorelines of Lake Russell.

> Prominent shore lines bordering Mono Lake that reach a height of 655 feet above the present water surface were cut in Late Pleistocene (Tioga) and Recent time by Lake Russell.
>
> The field evidence indicates that the high stand of the lake was essentially synchronous with the maximum advance of Tioga ice, and that, as the ice withdrew, the lake level was accordingly lowered. (Putnam, 1950, p. 121)

By showing that the lake was "essentially synchronous" with glaciers, Putnam helped overturn the belief that pluvial times (when desert lakes were large) alternated with glacial episodes, the lakes supposedly forming only after the glaciers melted. For the most part pluvial lakes and glaciers exist together, although they do not necessarily behave in exactly the same way. The maximum height of Mono Lake, reached 12,000–14,000 years ago, occurred several thousand years after the maximum development of the Tioga glaciers (Lajoie and Robinson, 1982; Bursik and Gillespie, 1993). This may indicate "that the increase in temperature at the end of the Pleistocene was not accompanied by a decrease in precipitation" (Lajoie and Robinson, 1982).

MOUNT LYELL AND VICINITY

Willard D. Johnson studied the landscape around Mount Lyell (in Yosemite National Park) and made valuable contributions about the origin of glacial landscapes and cirques. He drew special attention to the heads of glacial valleys:

> In ground plan, the canyon heads crowded upon the summit upland, frequently intersecting. They scalloped its borders, producing remnantal-table effects. . . . The summit upland—the preglacial upland beyond a doubt—was recognizable only in patches, long and narrow and irregular in plan, detached and variously disposed as to orientation, but always in sharp tabular relief and always scalloped. I likened it then, and by way of illustration I can best do so now, to the irregular remnants of a sheet of dough, on the biscuit board, after the biscuit tin has done its work. (Johnson, 1904, p. 571)

"Biscuit-board topography" has since been included in many textbooks and it is illustrated in figure 19 in the sketch that depicts the early stages of cirque development. Johnson showed how enlargement of cirques consumed preglacial topography and produced a variety of erosional landforms.

> Deep canyons had resulted, indirectly, because recession, directed horizontally, had been directed into a rising grade. This action seemed distinct from

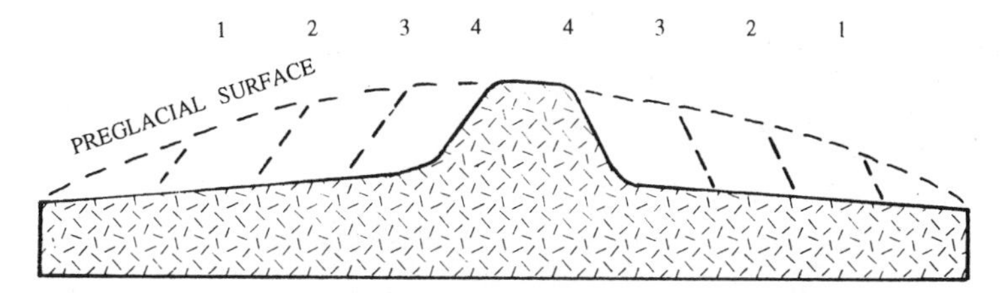

Figure 21. Basal sapping at the heads of glaciers causes cirques to develop, or "grow," headward into mountainsides. The numbers indicate positions of cirque headwalls at successive times. Destruction of the preglacial land surface leaves a mountain with a tabular summit. Continuation of the process will cause the cirque headwalls to intersect and form a pointed peak, or horn.

> that of abrasion. Abrasion accomplishes deepening vertically and directly. In the case of a "continental" glacier upon a level plain, abrasion would be operative alone. But that process was not to be invoked in explanation of the scalloped, tabular forms of the High Sierra; these pointed only to basal sapping.
> (Johnson, 1904, p. 572)

This is a clear statement of the difference between erosion of ice sheets and mountain glaciers, and the importance of basal sapping (fig. 21) in the development of glaciated mountains. Johnson suspected basal sapping occurred at the head of a glacier where ice, just thick enough to flow, began to move away from the mountainside, and he confirmed this by descending into a crevasse at the head of the Lyell Glacier.

JUNE LAKE LOOP

This loop is a popular and scenic semicircular road off of Highway 395 south of Mono Lake, serving the recreation and summer-home area of June, Gull, Silver, and Grant Lakes. A prominent landmark on the loop is a large erratic boulder above the road in the town of June Lake (fig. 22). An unusual feature of the area is Reversed Creek:

> June Lake, which lies at the eastern end of the horseshoe-shaped cañon we have described, and partially without the mountains, instead of discharging its drainage northward into Mono Valley, as would seem most natural, drains southward into the hills and is tributary to Rush Creek. The ancient drainage has been reversed by the deposition of morainal débris; we have therefore called the stream draining June and Gull lakes, Reversed Creek. The drainage before the site of June Lake was occupied by a glacier must have been northward.
> (Russell, 1889, p. 343)

Figure 22. Large glacial erratic balanced on a roche moutonnée in the town of June Lake, east of the Sierra Nevada.

Russell described how a 1,600-foot-thick glacier flowing down Rush Creek was split into two lobes by a resistant mountain and so created a horseshoe-shaped trough occupied by several lakes. The terminal moraine of the east lobe of the Rush Creek Glacier supposedly prevented June and Gull Lakes from draining north, so they drained south toward the mountains ("reversed"), eroding a channel to Silver Lake, and from there north to Mono Lake.

A more likely explanation of Reversed Creek is that of W. C. Putnam (1949). He mapped the bedrock of the area and showed that the June Lake part of the horseshoe valley is underlaid by more resistant rock than the western arm. Two streams once flowed close to each other out of the mountains; the western stream (Rush Creek) captured the eastern stream (Reversed Creek) by piracy along fractured rock of a fault zone. Then glaciation occurred, and the June Lake arm was eroded enough to change its slope from down-to-the-north to down-to-the-south. After the ice melted, Gull Lake drained south, "reversed," into Rush Creek. Thus glacier erosion rather than glacial deposition reversed the drainage.

SEQUOIA NATIONAL PARK

Sequoia National Park in the southern Sierra Nevada includes the highest peaks in the range. Road access is quite limited, and most of the area is visited on foot or horseback.

Figure 23. The nearly flat summit of Mount Whitney is a remnant of a gentle landscape that formed by erosion millions of years ago. Erosion by rivers plus multiple glaciation during the Ice Age has destroyed most of the ancient surface. The steep east face of Mount Whitney was produced by basal sapping at the heads of cirques. Numerous cirques, arêtes, and horns are visible in the distance. (Photo by F. E. Matthes, U.S. Geological Survey.)

The headwaters of the Kern River in the vicinity of Mount Whitney is rich in flat, eroded surfaces, which also occur at various elevations throughout the Sierra Nevada. The highest, most visible surfaces are the nearly flat summits of high peaks, often referred to as tabular summits. The highest peak in the Sierra Nevada, Mount Whitney, is not pointed but instead has a nearly flat summit several acres in extent (fig. 23).

> I grant you that in point of spectacular beauty Mount Whitney cannot compare with the dazzling Jungfrau or the sharp-profiled Matterhorn, but to one who can read it, the story told by its flat summit is far more significant and goes back to much more remote ages than any message conveyed by those glamorous alpine peaks. Indeed, the more fully I comprehend its story, by dint of repeated visits to and flights around and over Mount Whitney, the more venerable, the more precious seems that bit of flat land on its lofty summit. Upon it I have never set foot without a certain sense of reverence. (Matthes, 1937, p. 1)

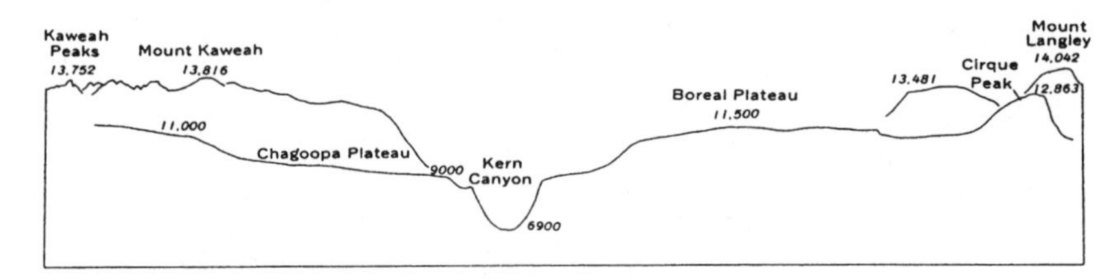

Figure 24. Preglacial erosion surfaces in the upper Kern River Basin are interpreted as former baselevels by Andrew Lawson and François Matthes. Current thinking is that these surfaces did not necessarily form at baselevel. (Drawing by U.S. Geological Survey, 2X vertical exaggeration; Matthes, 1950b.)

Andrew C. Lawson identified similar flat surfaces at lower elevations (fig. 24). He believed that each surface was eroded at a low elevation (baselevel) during a stable time and that uplift of the region raised it so that erosion could start over again to form a lower level (1904). His idea that such surfaces, found throughout the Sierra Nevada, formed at baselevel became accepted and was later important in efforts to understand the origin of Yosemite Valley.

Lawson mapped the glaciers that drained the upper Kern Basin and formed a trunk glacier that flowed through Kern Canyon (fig. 25) to south of Coyote Creek 28 miles from its source. From the amount of material in the terminal moraines, the depth of ice in the canyon, and other considerations, he concluded that the trunk glacier eroded only a few inches of rock from the walls of Kern Canyon:

> This estimate may be modified and corrected in a variety of ways, but it leaves a quantity for the glacial widening of the cañon which is negligible. The cañon then, had practically the same width before its occupancy by ice that it has today. (Lawson, 1904, p. 350)

Russell had arrived at a similar conclusion from his study of the ancient glaciers of the Mono Basin. Glaciers' effectiveness at erosion would be a major issue in the controversy about the origin of Yosemite Valley. Most geologists at the time believed glaciers could not significantly erode rock, and studies in California appeared to support that view. We now realize that the estimates Lawson and Russell made were incorrect because they ignored the large quantities of mud, silt, and sand (glacial outwash) that were carried away by rivers emanating from the glaciers.

Lawson agreed with W. D. Johnson regarding the origin of cirques and their importance in landscape development. He described how the growth of cirques can also cause drainage divides to shift sideways:

Figure 25. Looking south down Kern Canyon, a fine example of a U-shaped glacial valley in the southern Sierra Nevada. (Photo by F. E. Matthes, U.S. Geological Survey.)

> The recession of the eastern scarp of the Sierra Nevada by the sapping of glacial cirques has left Mt. Williamson an isolated stack, in the same sense that stacks are left upon a wave-cut terrace by the recession of the sea-cliff. The summit of this mountain is now a mile out from the crest line at Mt. Tyndall and half a mile out from the general line of the summit divide. It has been severed from the main mass of the mountains by cirque encroachment from the north. Lone Pine Peak, which stands as an isolated mass over two miles from the eastern scarp of the summit range, appears to be an even more striking illustration of the recession of that scarp. (Lawson, 1904, p. 360)

The westward shift of the Sierra crest because of glacial erosion creates the impression that Mount Whitney is lower than some lesser, but more easterly, peaks. This makes it difficult for some visitors to Whitney Portal, west of Lone Pine, to correctly identify Mount Whitney, highest peak in the conterminous United States.

Lawson had an interesting idea about mountain glaciation:

> In viewing from Mt. Whitney the revolution in the geomorphy of the High Mountain Zone which has been wrought by ice sculpture, and particularly by the gnawing of the cirques into the heart of the mass, one cannot but reflect that, had glacial conditions continued for twice as long as they actually did, or at most three times as long, the entire summit tract would have been obliterated, in the sense of being truncated to the level of the cirque floors. It is

> interesting to reflect further that this process of truncation, as it approaches completion, would not only remove the mountain tops, but thereby, also, do away with glaciation. Glaciation in the high mountains, in so far as it depends upon altitude, is, therefore, a process which automatically terminates.
>
> (Lawson, 1904, p. 360)

Between them Johnson and Lawson suggested a cycle of erosion by glaciers analogous to the erosion cycle of rivers, with its youthful, mature, and old age landscapes. The preglacial surface with extensive, gently sloped upland and small glaciers and cirques is youth; the deeply eroded landscape with large cirques and glacial troughs, horns, and arêtes is maturity; the subdued, stable landscape lying largely at the elevation of the floors of the cirques, with glaciers wasting, is old age. Uplift of the region would start the cycle over again.

F. Fryxell edited and published two works by François Matthes concerning Sequoia National Park. One is a collection of photographs with extended explanations that illustrate various geologic features, and browsing in it is a good way to learn Sierra Nevada geology (Matthes, 1950b). *A Glacial Reconnaissance of Sequoia National Park, California* includes a map showing the distribution of former glaciers, the southernmost glaciers of any size in the Sierra, and many fine photographs of glacial features (Matthes, 1965).

MCGEE MOUNTAIN: SUMMIT TILLS AND FAULTS

McGee Mountain is east of the Sierra crest between Bishop and Mammoth Lakes. The oldest Ice Age glacial deposit in California is found here (figs. 26, 27), and its location bears upon the history of the fault scarp along the east side of the Sierra Nevada:

> The truly noteworthy feature of McGee Mountain is that on its summit, and in the canyons marginal to the mountain, are preserved the most complete record of the glacial succession in the east-central Sierra Nevada. Part of this record is clearly visible to the most casual observer driving along the highway. This is the series of great lateral moraines nested inside one another at the mouths of the deep canyons flanking the mountain. These moraines are most conspicuously developed along Convict Creek, but they are impressive elements of the terrain at the mouth of McGee Canyon as well.
>
> Less apparent, but perhaps of greater scientific interest, are the discontinuous patches of till high up on the summit ridges of the mountain. The white boulders can be seen from afar, especially from the north, as a tattered blanket only partly covering the somber, reddish-brown, metasedimentary rocks on which they rest. These high-level, boulder-studded deposits are the McGee till. . . .
>
> (Putnam, 1962, p. 182)

Figure 26. McGee Mountain, east side of the Sierra Nevada. The arrow indicates the location of the McGee Till. Looking south from Highway 395.

Figure 27. Light-colored granitic boulders of McGee Till rest on darker-colored metamorphic rock high on McGee Mountain. Telephoto from near Convict Lake.

Because McGee Till consists largely of granitic materials resting on metamorphic rock it clearly has been transported. Blackwelder and Putnam give consideration to the possibility that the deposit is of mudflow origin, rather than glacial, but both conclude that it is indeed a glacial deposit (Blackwelder, 1931; Putnam, 1962). Curry reports similar "mountain summit glacial tills" 45 miles north of McGee Mountain in the area of Conway Summit and Yosemite National Park and says they are of "clear glacial origin" (1984).

Figure 28. Movement on a fault has displaced the crest of a lateral moraine about 50 feet. The line shows the approximate location of the fault, and arrows show the relative movement across the fault. McGee Creek Canyon, east side of the Sierra Nevada.

On McGee Mountain, a deep canyon now separates McGee Till from its source area, as much as 9 miles away at the Sierra crest, and the till is only found high on present-day divides, never in valleys. This unique distribution suggests there were no valleys east of the Sierra crest at the time of the McGee glaciation, about 1.5 million years ago. If there were no valleys, there could not have been a fault scarp. For glaciers to form, the mountains had to be close to their present elevation, so it appears that the composite fault scarp east of the Sierra Nevada formed by down-dropping of the land east of the Sierra crest rather than by uplift of the mountains. Matthes presents other lines of evidence in support of this view (1950a).

Fault scarps offset moraines of Tioga and Tahoe age at many places in the Owens Valley–Mono Lake area (fig. 28). Faults have displaced Tioga deposits as much as 26 meters and Tahoe deposits as much as 130 meters, with slip rates of between 0.4 and 1.3 millimeters per year for the last 10–25,000 years (Berry, 1990). Large earthquakes produced surface rupture in 1872 (Owens Valley) and 1980 (Hilton Creek), and small earthquakes occur every year.

PIRACY NEAR KEARSARGE PASS

Adolph Knopf found a rare example of glacier piracy where a glacier stole ice from a glacier in an adjacent valley:

> Westward from Kearsarge Pass a broad glacial valley extends, to all appearance, continuously westward. At Bullfrog Lake, however, it is found that the south wall of this seemingly continuous valley has been widely breached, and

> that the drainage, instead of flowing westward . . . escapes precipitously through this breach to Bubbs Creek, 1,100 feet below. . . . The lower part of the beheaded valley is drained by Charlotte Creek, and to ascend this stream is to gain an even more striking impression of the glacial capture of its upper part. For at the "elbow of capture" the valley floor descends abruptly to the level of Bullfrog Lake, and this downstepping of the valley floor upstream is of course in remarkable contrast to the normally stepped character of glacial valleys.
>
> (Knopf, 1918, p. 101)

After piracy, continued erosion by ice lowered the upper part of the valley 100 feet below the lower part of the valley, now abandoned, a circumstance that created the most unusual upstream-facing glacial step.

BRIDGEPORT AREA

Near Bridgeport, north of Mono Lake, are moraines of five major glacial advances and evidence of deformation of the region during the Ice Age. Sharp describes glaciers of Virginia Creek, Green Creek, Robinson Creek, and Buckeye Creek and includes maps and oblique aerial photographs that show the glacial features of the area; the annotated aerial photographs are especially instructive (1972).

Blackwelder describes interesting wind-abraded boulders on moraines near Bridgeport (1929). The sides of boulders facing glaciated canyons show abundant pits, grooves, and facets caused by sandblasting during the Ice Age. Similar boulders are found on moraines elsewhere east of the Sierra crest, but abrasion does not seem to be occurring in the present-day climate. The explanation is that when large glaciers filled the valleys they cooled the air in contact with the ice, increased its density, and created strong down-valley winds that today's empty valleys cannot duplicate.

LAKE TAHOE AREA

This popular area does not have the great elevation of the Sierra farther south, but there are well-developed glacial features in the Desolation Wilderness west of the lake and an interesting and unusual story of large floods caused by glaciers on the Truckee River north of Lake Tahoe.

After Yosemite, Joseph LeConte visited Lake Tahoe with his University Excursion Party and thought he saw evidence that the entire Tahoe Basin had been occupied by glaciers. We now know that the entire basin was never under ice, although some people still mistakenly believe that, like Yosemite Valley, the Tahoe Basin was formed by glacial erosion. Glaciers flowed into the basin from the west for short distances only and had no role in forming

Figure 29. Emerald Bay at Lake Tahoe is formed by an end moraine that almost completely isolates the bay from the main part of the lake.

it. (The Tahoe Basin was formed by faulting that caused a large slab of the earth to sink thousands of feet and leave steep mountain slopes on either side.) The most prominent glacial feature of Lake Tahoe is Emerald Bay, on the southwest shore, which is formed by a moraine loop that almost, but not quite, isolates the bay from the rest of the lake (fig. 29).

Jones studied the area south and west of Lake Tahoe and described the popular backpacking area of Desolation Valley as "filled to overflowing" and a "mer de glace" (sea of ice). His map shows moraines in Emerald Bay and around Cascade Lake and Fallen Leaf Lake. He saw the importance of glacial scenery to tourism, an enterprise that would flourish and dominate the future of the High Sierra:

> The wild and rugged scenery of the glaciated mountain zone constitutes the chief attraction for the many summer visitors to this general region. The land forms in themselves possess a beauty of most arresting character.
>
> (Jones, 1929, p. 139)

Peter Birkeland studied glacial deposits along the Truckee River north from Lake Tahoe and found evidence that glaciers had dammed the river and created great floods:

> Altitudes of Donner Lake Till and erratics in the Upper Truckee Canyon indicate that the ice was at least 840 feet thick near the mouth of Pole Creek and 440 feet thick at the north end of the canyon. Therefore the level of Lake

Tahoe may have risen to about 6,840 feet (a rise of about 600 feet) due to ice occupying the canyon. (Birkeland, 1964, p. 817)

Lakes dammed by glaciers are dangerous because the dam is unstable:

[S]ometimes annually, the lakes are rapidly emptied due to water initially escaping beneath the ice. This self-emptying is called a jökulhlaup in Iceland and is apparently initiated when the height of the dammed lake is about 9/10 that of the ice barrier. When these conditions are reached the water buoys up the adjacent ice and then escapes by forcing its way downvalley along the base of the glacier. The water emerges at the glacier snout in such volumes that catastrophic floods are common downstream.

Jökulhlaups could have taken place at numerous times during any of the glaciations whenever canyon ice ponded the drainage and thereby caused the level of Lake Tahoe to rise . . . [Resulting] . . . floods may help to account for the large size of the granitic boulders in the glacial outwash, even as far downstream as Reno, Nevada, where 10-foot boulders have been observed.

(Birkeland, 1964, p. 819)

Glaciers were larger west of Lake Tahoe than to the east:

Huge valley glaciers moved down the canyons along the western side of the lake scouring away all of the loose rock and built up great piles of morainal debris. During their period of maximum development, these ice streams were as much as 1,000 feet thick and, in some areas, covered all but the highest peaks and ridges. Along the eastern side of the basin, glaciers developed only on the shaded sides of the highest peaks so most of this area was not glaciated. This change from west to east across the basin accounts for the rugged alpine scenery found along the Sierran crest and the subdued rolling topography typical of most of the Carson Range. (Burnett, 1971, p 121)

Mountains east of the lake receive little precipitation because they are in the rain shadow of the Sierra Nevada.

MOKELUMNE, YUBA, AND FEATHER RIVERS

The northern Sierra Nevada is not high in elevation, but there have been large glaciers there nonetheless:

Field reconnaissance and aerial photographic analysis of the central Sierra Nevada demonstrate that the Mokelumne and Carson River drainages were occupied by a large late-Pleistocene ice field. This ice field covered more of the range crest than did any glacier or ice field to the south. Reconnaissance investigations further north suggest that the Yuba and possibly the Feather

River drainages also supported large ice fields. Our study thus challenges the concept that large ice fields with substantial transfluent flow across the Sierra crest did not exist in the northern half of the range.

(Clark, Clark, and Gillespie, 1991, p. 13)

Farther north in the vicinity of the Feather River,

glaciers occurred all along the crest as far north as the headwaters of Butt Creek. In the Diamond Mountains, small glaciers were present on the north slope of the north end of Grizzly Ridge overlooking Indian and Genesee valleys, on Keddie Peak, Kettle Rock, and on the north end of the Honey Lake fault scarp overlooking Janesville and Susanville. On the east side of Mt. Ingalls, a 3-mile-long glacier occupied the valley of Coldwater Creek.

(Durrell, 1986, p. 277)

Butt Creek is at the northwest end of the Sierra Nevada close to Lassen Volcanic National Park, where rock typical of the Sierra Nevada is buried by young volcanic rocks of the Cascade Mountains.

The readily accessible and pleasant Sierra Buttes–Lakes Basin Recreation Area between the Middle Fork of the Feather River and the North Fork of the Yuba River contains well-developed glacial features. The Sierra Buttes, 8,587 feet elevation, has a fine cirque on its east side from which extends a glacial valley with several lakes, glacial steps, and large lateral moraines.

REFERENCES CITED

Alpha, Tau Rho, Clyde Wahrhaftig, and N. King Huber. 1987. *Oblique map showing maximum extent of 20,000-year-old (Tioga) glaciers, Yosemite National Park, California.* U.S. Geol. Survey Misc. Investigations Series Map I-1885.

Bateman, P. C., R. W. Kistler, D. L. Peck, and A. J. Busacca. 1983. *Geologic map of the Tuolumne Meadows quadrangle, Yosemite National Park, California.* U.S. Geol. Survey Geologic Quadrangle Map GQ-1570, scale 1:62,500.

Berry, M. E. 1990. Soil-geomorphic analysis of late Quaternary glaciation and faulting, eastern escarpment of the central Sierra Nevada, California. Ph.D. diss., Univ. of Colorado, Boulder.

Birkeland, P. W. 1964. Pleistocene glaciation of the northern Sierra Nevada north of Lake Tahoe. *Journal of Geology* 72: 810–25.

Blackwelder, Eliot. 1929. Sandblast action in relation to the glaciers of the Sierra Nevada. *Journal of Geology* 37: 256–60.

———. 1931. Pleistocene glaciation in the Sierra Nevada and Basin Ranges. *Geol. Soc. of Amer. Bull.* 42: 865–922.

Burnett, John L. 1971. Geology of the Lake Tahoe basin. *Calif. Geology* 24: 119–27.

Bursik, M. I. 1988. Late Quaternary volcano-tectonic evolution of the Mono Basin, eastern California. Ph.D. diss., Calif. Institute of Technology.

Bursik, M. I., and A. R. Gillespie. 1993. Late Pleistocene glaciation of Mono Basin, California. *Quaternary Research* 39: 24–35.

Clark, Douglas H., Malcolm M. Clark, and Alan R. Gillespie. 1991. A Late Pleistocene ice field in the Mokelumne drainage, north-central Sierra Nevada. *Geol. Soc. of Amer. Abstracts with Programs* 23, no. 2: 13.

Clark, M. M. 1976. Evidence for rapid destruction of latest Pleistocene glaciers of the Sierra Nevada, California. *Geol. Soc. of Amer. Abstracts with Programs* 8: 361–62.

Curry, Robert R. 1984. Mountain summit glacial tills and their tectonic implications, eastern Sierra Nevada, California. *Geol. Soc. of Amer. Abstracts with Programs* 16: 481.

Durrell, Cordell. 1986. *Geologic history of the Feather River country.* Berkeley: Univ. of Calif. Press.

Johnson, W. D. 1904. The profile of maturity in alpine glacial erosion. *Journal of Geology* 12: 569–78.

Jones, Wellington D. 1929. Glacial landforms in the Sierra Nevada south of Lake Tahoe. *Univ. of Calif. Publications in Geography* 3: 135–57.

Knopf, Adolph. 1918. *A geologic reconnaissance of the Inyo Range and the eastern slope of the southern Sierra Nevada, California.* U.S. Geol. Survey Prof. Paper 110.

Lajoie, K. R. 1968. Late Quaternary stratigraphy and geologic history of Mono Basin, eastern California. Ph.D. diss., Univ. of Calif. at Berkeley.

Lajoie, K. R., and S. W. Robinson. 1982. Late Quaternary glacio-lacustrine chronology, Mono Basin, California. *Geol. Soc. of Amer. Abstracts with Programs* 14: 179.

Lawson, Andrew C. 1904. The geomorphogeny of the upper Kern Basin. *Univ. of Calif. Publications, Bulletin of the Dept. of Geology* 3: 291–376.

LeConte, Joseph. [1870; revised 1900] 1971. *A journal of ramblings through the High Sierra of California.* Reprint, New York: Sierra Club/Ballantine Books.

Matthes, François E. 1937. The geologic history of Mount Whitney. *Sierra Club Bulletin* 22: 1–18.

———. 1950a. *The incomparable valley: A geological interpretation of the Yosemite.* Edited by F. Fryxell. Berkeley: Univ. of Calif. Press.

———. 1950b. *Sequoia National Park: A geological album.* Edited by F. Fryxell. Berkeley: Univ. of Calif. Press.

———. 1965. *Glacial reconnaissance of Sequoia National Park, California.* Edited by F. Fryxell. U.S. Geol. Survey Prof. Paper 504A.

Muir, John. [1912] 1962. *The Yosemite.* Garden City, N.Y.: Anchor Books.

Putnam, William C. 1949. Quaternary geology of the June Lake district, California. *Geol. Soc. of Amer. Bull.* 60: 1281–302.

———. 1950. Moraine and shoreline relationships at Mono Lake, California. *Geol. Soc. of Amer. Bull.* 61: 115–22.

———. 1962. Late Cenozoic geology of McGee Mountain, Mono County, California. *Univ. of Calif. Publications in Geological Sciences* 40: 181–218.

Russell, Israel C. 1889. Quaternary history of Mono Valley, California. U.S. Geol. Survey 8th Annual Report, 261–394.

Sharp, Robert P. 1972. Pleistocene glaciation, Bridgeport Basin, California. *Geol. Soc. of Amer. Bull.* 83: 2233–60.

Whitney, J. D. 1865. The High Sierra. Chap. 10 in *Report of progress, and synopsis of field work from 1860 to 1864.* Vol. 1 of *Geology.* Philadelphia: Geol. Survey of California.

CHAPTER FOUR

ORIGIN OF YOSEMITE VALLEY

THE YOSEMITE PROBLEM

Yosemite Valley deserves a chapter of its own because it is one of the world's special places, with an interesting geologic story as well as an interesting human history. This chapter summarizes the development of ideas about the origin of Yosemite Valley and provides an example of how science progresses by the accumulation of knowledge by many different people. Because they aid historical understanding, some quotes are included even though they contain statements that are outdated; beware of taking quotes out of context.

Yosemite Valley is famous for its beauty, and tourism has thrived there since nonnatives "discovered" it in 1851 (fig. 30). What makes Yosemite Valley special is the pleasing juxtaposition of cliffs, waterfalls, and domes with a flat valley floor over which a lovely river meanders through wood and meadow. Visitors are naturally curious about the origin of such a magnificent place.

Efforts to understand the origin of Yosemite Valley began in the 1860s with study by the California Geological Survey. In the 1870s the origin of the valley became the subject of a dispute, sometimes bitter, known as the "Yosemite Problem" or the "Yosemite Controversy." At first, debate centered upon the competing roles of faulting, advocated by Josiah D. Whitney, and glacial erosion, advocated by John Muir. Later study and debate concerned the relative importance of erosion by glacier and by river. In 1913 the U.S. Geological Survey assigned François Matthes the task of determining the valley's origin. Upon publication in 1930 his work was well received, and it ended the Yosemite Controversy. Knowledge gained since 1930 requires modification of some of Matthes's work, and these new ideas are briefly pre-

Figure 30. Yosemite Valley, looking east. The flat, forested, floor of the valley was formed by in-filling of a lake. Thick lake sediments mask the U-shape of the glacial valley.

sented in this chapter. Finally a summary of the geologic development of Yosemite Valley is given.

THE BOTTOM DROPPED OUT (JOSIAH WHITNEY)

Josiah D. Whitney thought Yosemite Valley was formed by down-dropping of a large block of rock between faults, and that glaciers were of no importance at all. This hypothesis had little evidence to support it, but it is interesting to examine as an example of how appearances can deceive.

Features of Yosemite Valley that most impressed Josiah D. Whitney were the steep, nearly parallel valley walls, with little talus at their bases, the sheer cutoff face of Half Dome, and the flat valley floor (fig. 31). Whitney knew these forms were not typical of glacial valleys in the Alps. The narrow, V-shaped gorge that exists along Tenaya Creek above Yosemite also argued against a glacial origin and suggested some special, local, process had been at work. Of Half Dome he wrote:

> It is an inaccessible crest of granite, rising to the height of 4737 feet above the valley, the face fronting towards Tenaya Creek being *absolutely vertical* for 2000 feet down from the summit. . . . [I]t is an unique thing in mountain scenery, and nothing even approaching it can be found except in the Sierra Nevada itself.
> (Whitney, 1865, p. 416)

Figure 31. The Merced River in Yosemite Valley meanders on the nearly flat surface of in-filled Lake Yosemite.

Whitney concluded that the unique forms of Yosemite Valley must have formed by down-dropping between faults:

> [I]t appears to us probable that this mighty chasm has been roughly hewn into its present form by the same kind of forces which have raised the crest of the Sierra. . . . The Half Dome seems, beyond a doubt, to have been split asunder in the middle, the lost half having gone down in what may truly be said to have been "the wreck of matter and the crush of worlds."
>
> (Whitney, 1865, p. 421)

> We conceive that, during the process of upheaval of the Sierra or, possibly at some time after that had taken place, there was at the Yosemite a subsidence of a limited area, marked by lines of "fault" or fissures crossing each other somewhat nearly at right angles. In other and more simple language, the bottom of the Valley sank down to an unknown depth. . . .
>
> (Whitney, 1868, p. 77)

> Lake Tahoe and the valley which it partly occupies we conceive also to be, like the Yosemite, the result of local subsidence. (Whitney, 1868, p. 79)

Clarence King, who was familiar with glacial features in Yosemite Valley, thought Whitney was right, as did the prominent geologists Israel C. Russell and Andrew Lawson, who, from studies elsewhere in the Sierra, had con-

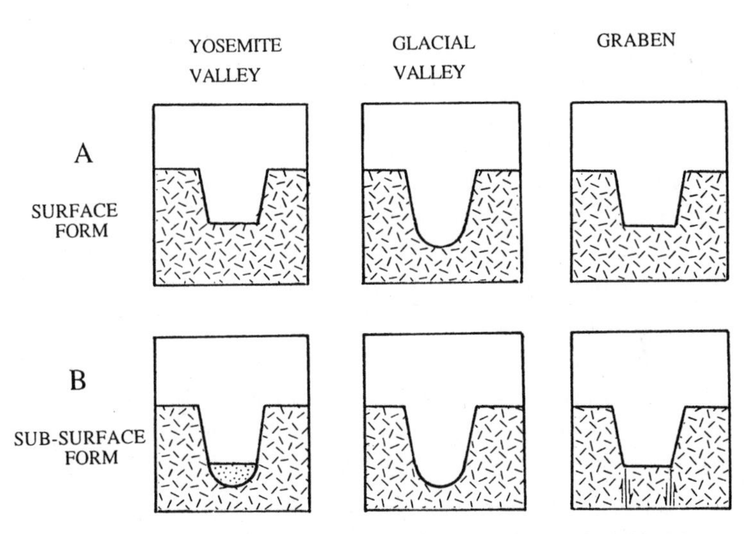

Figure 32. A. In basic surface form, Yosemite Valley resembles a fault-block basin, or graben, more than it does a glacial valley. B. Beneath the thick lake sediments is the rest of a U-shaped glacial valley, not a block of rock that has dropped down between faults.

cluded that glaciers did little eroding of the valleys through which they flowed.

Whitney knew some glacial features existed in the valley and that one moraine in particular

> may have acted as a dam to retain the water within the valley, after the glacier had retreated to its upper end, and that it was while thus occupied by a lake that it was filled up with the comminuted materials arising from the grinding of the glaciers above, thus giving it its present nearly level surface.
>
> (Whitney, 1865, p. 423)

Writing about the lake, Whitney came close to grasping an important point that, if understood, might have caused him to recognize that glaciers were important in creating Yosemite Valley, thereby avoiding the Yosemite Problem, and might have caused history to be kinder to him. But nature tricked Whitney, as illustrated by figure 32. Thick lake deposits obscure an important part of the glacial evidence, the bottom of the U-shaped valley. What is left to be seen of the U, together with the flat valley floor formed by the lake deposits, is a good imitation of a down-dropped fault block, or graben. If lake sediments were not present, the U-shape of the underlying bedrock would be visible, as in many other Sierra valleys, Yosemite Valley would not resemble a graben, and Whitney would probably have recognized the importance of glaciers in its development.

But how did Whitney miss other signs that point to the importance of glaciers? After all, is not Yosemite the finest example of a glacial valley anywhere? No, it is not, and this is a surprise for many people. There is much to praise about Yosemite Valley, but to praise it, as has been done, as a "sublime monument to glaciers" or as "nature's textbook on glacial erosion" is misleading. Many valleys in the Sierra Nevada display more characteristic glacial features than Yosemite Valley does. Yosemite abounds in unusual landforms, features not at all typical of glaciated valleys but which are more conspicuous than the glacial landforms present. The steep cliffs and somewhat straight, open valley are of glacier origin, but they are not uniquely so, because other geologic processes create similar features. Moraines and small areas of glacial polish are present but are not prominent, and most visitors neither notice nor photograph them.

By contrast, the outstanding landforms of Yosemite Valley, the trademark features that are photographed, gawked at, climbed on, painted, hiked to, admired, and that adorn T-shirts, book covers, and corporate logos are neither characteristic of, nor unique to, glacial erosion. Half Dome owes its special form to weathering, exfoliation, and a vertical joint. North Dome and Sentinel Dome formed by weathering and exfoliation, not glacial erosion. El Capitan is the imposing monument it is because the rock of which it is composed has few joints; its near-verticality is in part attributable to glacial erosion, but imposing cliffs are not unique to glacier erosion. Distinctive forms such as Royal Arches, Washington Column, and Three Brothers were shaped by jointing of diverse orientation and form, plus removal of rock by rockfall or glacier plucking, and it is not clear which process removed any given block of rock. The flat valley floor was produced by in-filling of a lake; yes, the lake was of glacial origin, but, once again, such features are not unique to glaciated areas. What is surprising is not that Whitney failed to see the hand of glaciers, but that Muir and others did!

GLACIERS EXCAVATED IT (JOHN MUIR)

John Muir became champion of the idea that glaciers were responsible for excavating Yosemite Valley, but it should be noted that he was not the first to observe glacial features of the valley. In October of 1864, four years before Muir arrived in California, Clarence King of the California Geological Survey studied Yosemite Valley:

> I gathered ample evidence that a broad sheet of glacier, partly derived from Mount Hoffmann and in part from the Mount Watkins ridge and Cathedral Peak, but mainly from the great Tuolumne glacier, gathered and flowed into the Yosemite Valley. Where it moved over the cliffs there are well-preserved

> scarrings. The facts which attest this are open to observation, and seem to me important in making up a statement of past conditions. (King, 1872, p. 165)

> Up the Yosemite gorge, which opened straight before me, I knew that another great glacier had flowed; and also that the valley of the Illilouette and the Little Yosemite had been the bed of rivers of ice; a study, too, of the markings upon the glacier cliff above Hutching's house, had convinced me that a glacier no less than a thousand feet deep had flowed through the valley, occupying its entire bottom. (King, 1872, p. 168)

King did not have opportunity to follow up on his early study of Yosemite Valley.

When John Muir first visited Yosemite in 1868 at age thirty he lacked formal credentials as a scientist, but he had attended a university for several years, was knowledgeable about many aspects of natural history, and was a careful and perceptive observer. He lived in Yosemite Valley for many years and had abundant opportunity to study glacial features and reflect on their significance. Muir saw no merit in the subsidence idea of Whitney:

> The argument advanced to support this view is substantially as follows: it is too wide for a water-eroded valley, too irregular for a fissure valley, and too angular and local for a primary valley originating in a fold of the mountain surface during the process of upheaval; therefore a portion of the mountain bottom must have suddenly fallen out, letting the super-incumbent domes and peaks fall rumbling into the abyss, like coal into the bunker of a ship. This violent hypothesis, which furnishes a kind of Tophet for the reception of bad mountains, commends itself to the favor of many by seeming to account for the remarkable sheerness and angularity of the walls, and by its marvelousness and obscurity, calling for no investigation, but rather discouraging it. (Muir, in Colby, 1950, p. 18)

"Tophet" refers to a shrine near Jerusalem where human sacrifice was once practiced. Lack of an academic degree was no handicap to Muir; he was a formidable opponent to Whitney and the better writer, one who was later amazed that publishers would pay him money for his writings about Yosemite and the High Sierra.

Muir was no half-hearted glacier booster. He was certain that glaciers alone had excavated Yosemite Valley, that all of California was once covered by a great ice sheet (see chapter 2), and that throughout the Sierra Nevada, glaciers alone had excavated every valley and peak:

> In the development of these, the Master Builder chose for a tool, not the earthquake nor lightning to rend and split asunder, not the stormy torrent nor eroding rain, but the tender snow-flowers noiselessly falling through unnumbered seasons, the offspring of the sun and sea. If we should attempt to restore the range to its pre-glacial unsculptured condition, its network of profound cañons would have to be filled up, together with all its lake and meadow basins; and every rock and peak, however lofty, would have to be buried again beneath the fragments which the glaciers have broken off and carried away.
> (Muir, in Colby, 1950, p. 3)

Joseph LeConte of the University of California in Berkeley sought out Muir in 1870 to discuss the origin of Yosemite Valley, and wrote,

> He fully agrees with me that the peculiar cleavage of the rock is a most important point, which must not be left out of account. He further believes that the valley has been wholly formed by causes still in operation in the Sierra—that the Merced Glacier and the Merced River and its branches, when we take into consideration the peculiar cleavage, and also the rapidity with which the fallen and falling boulders from the cliffs are disintegrated into dust, have done the whole work. The perpendicularity is the result of cleavage; the want of talus is the result of the rapidity of disintegration and the recency of the disappearance of the glacier. I differ with him only in attributing far more to preglacial action. (LeConte, 1870, p. 68)

Cleavage is what, today, we call *joints* (fig. 33). Joints are fractures in rock produced by stress, such as occurs during uplift of a mountain range, release of pressure after removal of overlying rock, or cooling of a lava flow. Joints hasten destruction of rock by permitting entry of water, air, and acids from decomposing vegetation. In his article "Mountain Sculpture," Muir shows that he understood the importance of joints in erosion, though he often single-mindedly went ahead and stressed glaciers to the exclusion of other geologic factors (Colby, 1950).

Muir traced the pathway of glaciers that converged on Yosemite Valley:

> They number five, and may well be called Yosemite glaciers, since they were the agents by which beauty-loving nature created the grand valley, grinding and fashioning it out of the solid flank of the range, block by block, particle by particle, with sublime deliberation and repose.
>
> The names I have given them are, beginning with the northmost, Yosemite Creek, Hoffman, Tenaya, South Lyell, and Illilouette [*sic*] Glaciers. They all converged in admirable poise around from north-east to south-east, welding themselves together into one huge trunk which swept down through the valley, filling it brimful from end to end, receiving small tributaries on its way

Figure 33. Joints are cracks in rock that make rock more easily weathered and eroded by glaciers. Here, water freezing in joints has disrupted the rock. (Photo by F. E. Matthes, U.S. Geological Survey.)

> from the Indian, Sentinel, and Pohono Cañons; and at length flowed out of the valley, and on down the range, in a general westerly direction.
>
> (Muir, 1880, p. 553)

Muir happily undertook the task of educating the public about glaciers and was successful because of writing like this:

> The work of glaciers, especially the part they have played in sculpturing the face of the earth, is as yet but little understood, because they have so few loving observers willing to remain with them long enough to appreciate them. Water rivers work openly where people dwell, and so does the rain and the dew, and the great salt sea embracing all the world; and even the universal ocean of air, though invisible, yet it speaks aloud in a thousand voices, and explains its modes of working and its power. But glaciers, back in their cold solitudes, work apart from men, exerting their tremendous energies in silence and darkness. Outspread, spirit-like, they brood above the long predestined landscapes, working on unwearied through unmeasured ages, until, in the fullness of time, the mountains and valleys and plains are brought forth, channels furrowed for the rivers, basins made for the lakes and meadows and long, deep arms of the sea, soils spread for the forests and the fields—then they shrink and vanish like summer clouds. (Muir, 1880, p. 557)

No writer combines science and poetry better than John Muir, and this in part explains why his readers came to love him and believe in his ideas. Although he was overzealous about the erosional ability of glaciers, their former extent, and their effects on California and the Sierra Nevada, Muir focused attention on the importance of glaciers in eroding landforms and won praise from François Matthes:

> To one thoroughly at home in the geologic problems of the Yosemite region it is now certain, upon reading Muir's letters and other writings, that he was more intimately familiar with the facts on the ground and was more nearly right in their interpretation, than any professional geologist of his time.
>
> (Matthes, 1938, p. 10)

A DIGRESSION IN DEFENSE OF WHITNEY

Most historical accounts present Whitney unfavorably because he was wrong about the origin of Yosemite Valley and ill-mannered and arrogant toward John Muir, a popular underdog. The two men differed in all respects. Whitney was a professional geologist, graduate of a fine university, a respected scientist, and a responsible state geologist. Muir was a good-hearted vagabond, a nature lover, not a geologist or even a college graduate, and he did not claim to be a professional scientist. Whitney spent much time and energy managing a survey and fighting with the state legislature for its survival, while Muir lived carefree in Yosemite Valley, observing and thinking about the valley every day. A century later we would have said Whitney was establishment and Muir was a hippie. That such a person might challenge his geologic ideas, and be successful in so doing, infuriated Whitney. Frustrated and hurt by the popularity of Muir and his eroding ice, Whitney wrote harshly and condescendingly about him, calling him an "ignorant sheepherder" with "absurd" ideas, among other things. In his zeal to combat the glacier menace he even repudiated his friend Clarence King and his own earlier writing by denying that ice had ever occupied Yosemite Valley!

A myth has developed to the effect that Whitney was a pompous windbag who did nothing right while Muir was a neat guy who did nothing wrong. This is unfortunate and unfair. Whitney made major contributions to understanding Yosemite, the Sierra Nevada, and California, and he played an important role in getting Yosemite Valley preserved as public land. He had reason to hypothesize faulting in Yosemite Valley and the Tahoe Basin, was right about Tahoe, and correctly maintained that Muir had some wrong ideas about glaciers, though this latter fact is often discreetly overlooked. It does not detract from Muir for us to treat Whitney with more respect.

A. Glacial valley and moraines, Green Creek, east side of the Sierra Nevada. Photo by John S. Shelton.

B. Big Arroyo in Sequoia National Park is a good example of a glacial valley.

Plate 1. Glaciers existed in six geologic provinces of California during the Ice Age. The Sierra Nevada had the largest and most widespread glaciers. Ice Age glaciers were all gone by 10,000 years ago, leaving behind interesting and scenic glacial landforms.

A. Looking east from the summit of Sierra Buttes, northern Sierra Nevada. First there is a cirque lake, then a glacial step, then two more glacial lakes between large lateral moraines.

C. Glacial step with a cascading stream. Fourth Recess of Mono Creek, Sierra Nevada.

B. Glacial lakes occupy basins excavated by glaciers into the granitic rock of the Sierra Nevada. View from the summit of Mount Lyell, Yosemite National Park.

Plate 2. Ice Age glaciers removed soil and weathered rock, and left large areas of bare rock, thousands of lakes, glacial steps with waterfalls or cascades, and large moraines.

Above: A. Nested lateral moraines along McGee Creek, east side of the Sierra Nevada. Photo by D. D. Trent.

Left: B. Two lateral moraines and an end moraine preserve the outline of a former glacier in McGee Creek Canyon. An active fault, of which there are many along the east side of the Sierra Nevada, has offset the lateral moraines as much as 50 feet. Photo by John S. Shelton.

Plate 3. The glacial history of California is largely recorded in deposits of till, which often occur in ridges, called moraines. Moraines that formed at the margins of glaciers often preserve an outline of the glacier.

A. Mount Shasta, viewed from the north. The Whitney Glacier, largest in California, extends down to the right, beginning just to the right of the summit.

B. Crevasses in the Whitney Glacier, Mount Shasta.

Plate 4. New glaciers have formed in California in the last 700 years. Among them are the 9 glaciers and 1 glacieret that exist high on Mount Shasta in northern California.

A. Cirque with small unnamed glacier on the Silver Divide, Sierra Nevada.

B. East Sill Glacier and Norman Clyde Glacier (in the distance). Photo by Carl A. Yeary.

Plate 5. The 99 glaciers and 398 glacierets of the Sierra Nevada are located in cirques created by Ice Age glaciers. Bare ice is exposed in the lower parts of the glaciers by late summer.

Above: A. Aerial view of Palisades glaciers, Sierra Nevada. Bergschrunds are present on both glaciers. Photo by John S. Shelton.

Right: B. A crevasse in the Palisade Glacier. The boulder fell from a nearby cliff. Photo by D. D. Trent.

Plate 6. The Palisades area along the Sierra crest west of Big Pine has 10 glaciers and glacierets, including the Palisade Glacier, largest in the Sierra Nevada.

A. High peaks and moraines around the Palisade Glacier. Photo by D. D. Trent.

C. Climbing at the head of the Palisade Glacier. Photo by D. D. Trent.

B. Bergschrund and couloir at the head of the Palisade Glacier. Photo by D. D. Trent.

Plate 7. The Palisade Glacier is between 12,000 and 13,400 feet elevation on the northeast slope of North Palisade in an area of peaks over 14,000 feet high.

Above: A. Ice-cored end moraine, Palisade Glacier. Photo by D. D. Trent.

Right: B. Rock glacier in a cirque, Tinemaha Creek, Sierra Nevada. Photo by Douglas H. Clark.

Plate 8. California glaciers are smaller than when they were discovered in the 1870s. End moraines of some glaciers cover and protect glacier ice, which causes the moraines to be impermeable and to impound meltwater ponds in summer. A closely related landform is the rock glacier, some of which are glaciers covered with rock debris.

BOTH RIVER AND GLACIER SHAPED THE VALLEY (FRANÇOIS MATTHES)

The accepted version of the origin of Yosemite Valley gives somewhat equal importance to river and glacier in *deepening* the valley, while glacier erosion is regarded as the primary agent for *widening* the valley and creating the steep cliffs. In large measure, though not completely, it supports the ideas of John Muir.

In 1930 the United States Geological Survey published Professional Paper 160, *Geologic History of the Yosemite Valley*, based on years of field study by François Matthes and others. This well-written and illustrated work has many fine photographs and drawings that have been reproduced in hundreds of publications throughout the world. It is out of print, which is regrettable even though some aspects of the book are out of date.

Matthes wrote about the ideas of Whitney and Muir:

> The underlying idea of Whitney's hypothesis was, it should be added in all fairness, not so absurd as some of his opponents have intimated, for there are many well-authenticated instances of valleys that have been created by the subsidence of blocks of the earth's crust. Several such depressions are associated with the Sierra Nevada—notably Owens Valley and the basin occupied by Lake Tahoe. However Whitney's hypothesis rested on no tangible evidence, and nothing has been found thus far to substantiate it.
>
> (Matthes, 1930, p. 4)

> Muir, on his part, went too far in his claims for glacial erosion. . . . [H]e maintained that the Yosemite and, indeed, all the great canyons of the range, thousands of feet in depth, had been gouged out entirely by the glaciers.
>
> (Matthes, 1930, p. 5)

Matthes saw the Yosemite Problem not as a debate about faults and glaciers, but about the relative effects of glacier and river. He was able to make a better study than was possible previously because a new, detailed, topographic map of the valley had recently been prepared (by Matthes himself, in his role as topographer), and because of new ideas about how landscapes develop:

> [S]ince the end of the nineteenth century there had come to maturity a new branch of geologic science—geomorphology, or physiography, as it is also termed—which deals specially with the origin and development of the surface features of the earth and within whose scope a problem such as that of the Yosemite Valley largely belongs. (Matthes, 1930, p. 6)

The new ideas were mostly those of William Morris Davis, often regarded as the founder of geomorphology. Armed with a good map, new ideas, and lots of time, Matthes took to the field.

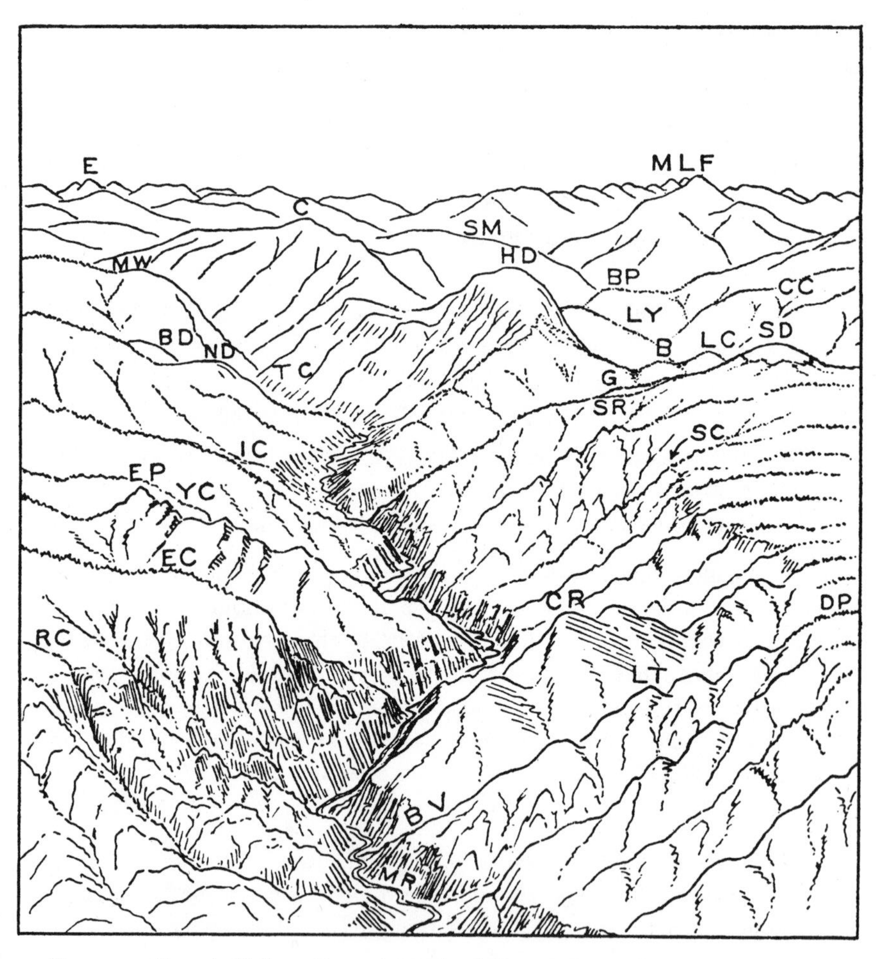

Figure 34. Yosemite Valley as François Matthes believes it appeared during his hypothetical Canyon Stage, just before the first Ice Age glaciation. (Drawing by U.S. Geological Survey; Matthes, 1930.)

THE DEPTH OF YOSEMITE VALLEY

Matthes undertook to determine the relative amounts of river erosion and glacier erosion in sculpturing Yosemite Valley. The result permits visualization of what Matthes believed Yosemite Valley looked like before and after glaciation (figs. 34, 35). Matthes calculated that, just before the Ice Age, the river-eroded canyon was about 2,000 feet deep below Glacier Point and about 2,400 feet deep below El Capitan. He estimated the amount of glacier erosion as

> the depth of glacial excavation decreases from a maximum of about 1,500 feet at the head to a minimum of about 500 feet at the lower end. The reason for

Figure 35. Yosemite Valley as François Matthes believes it appeared immediately after the last Ice Age glacier withdrew, perhaps 15,000 years ago. Multiple glaciation widened, deepened, and straightened the former river valley. Lake Yosemite, in part dammed by moraines, may have been several miles long but was not very deep. (Drawing by U.S. Geological Survey; Matthes, 1930.)

RC Ribbon Creek
EC El Capitan
EP Eagle Peak
YC Yosemite Creek
IC Indian Creek
R Royal Arches
W Washington Column
TC Tenaya Creek
ND North Dome
BD Basket Dome
MW Mount Watkins
E Echo Peak
C Clouds Rest
SM Sunrise Mountain
HD Half Dome
M Mount Maclure
L Mount Lyell
F Mount Florence
BP Bunnell Point
LY Little Yosemite Valley
B Mount Broderick
LC Liberty Cap
SD Sentinel Dome
G Glacier Point
SR Sentinel Rock
SC Sentinel Creek
CR Cathedral Rocks
BV Bridalveil Creek
LT Leaning Tower
DP Dewey Point
MR Merced River

> this decrease in glacial deepening down the valley is found in the fact that during all phases of glaciation the ice was thicker and therefore had greater excavating power at the head of the valley than at the lower end, and during the maximum phases it plunged into the head of the valley in the form of a mighty cataract. (Matthes, 1930, p. 102)

These numerical estimates are not as highly regarded today as they were in 1930, but the important conclusion endures: that river and glacier made somewhat equal contributions to the depth of Yosemite Valley.

How Matthes estimated the amount of downcutting done by the Merced River prior to glaciation is an interesting application of the geomorphic ideas of William Morris Davis. Even though the methodology is imperfect and based on some questionable assumptions, it will be summarized here because it is enlightening and of historic interest.

As the Sierra Nevada was being uplifted during millions of years prior to the Ice Age it was also tilting down to the west. In response, the Merced River deepened its valley, and did so faster than its tributary streams could, because, first, it was the larger stream and, second, it flowed west down the increasing slope of the range while most tributaries flowed north or south and were not directly invigorated by the westerly tilting. Matthes, following Davis, believed that mountain ranges were typically uplifted in jerks, with short periods of relatively rapid uplift followed by longer periods of stability. After each uplift the Merced River (supposedly) cut down to a new level of equilibrium, or baselevel, and a new landscape developed. The less powerful tributaries, unable to keep up with the rapid downcutting of the Merced, were left "hanging" above the master stream (fig. 36). The old slopes of the hanging tributaries still "point" to the former elevations of the Merced River (fig. 37).

By drawing profiles of numerous tributaries, Matthes re-created steps in the downcutting of the Merced River, as shown in figure 38. He named the reconstructed landscapes the "Broad-Valley," the "Mountain-Valley," and the "Canyon" Stages, each landscape supposedly being formed during a period of stability between uplifts. Matthes spoke of the Merced River as having a "three-story" profile, illustrated in his "Canyon Stage" of figure 38: each story, or form, corresponds (Matthes believed) to a single uplift and subsequent erosional episode. The erosion surfaces in the vicinity of Mount Whitney (chapter 3) appear to support this part of Matthes's work, but the support may not be valid, as will be explained later.

Figure 36. The valley of Bridalveil Creek "hangs" several hundred feet above the Merced River and the floor of Yosemite Valley, creating Bridalveil Fall.

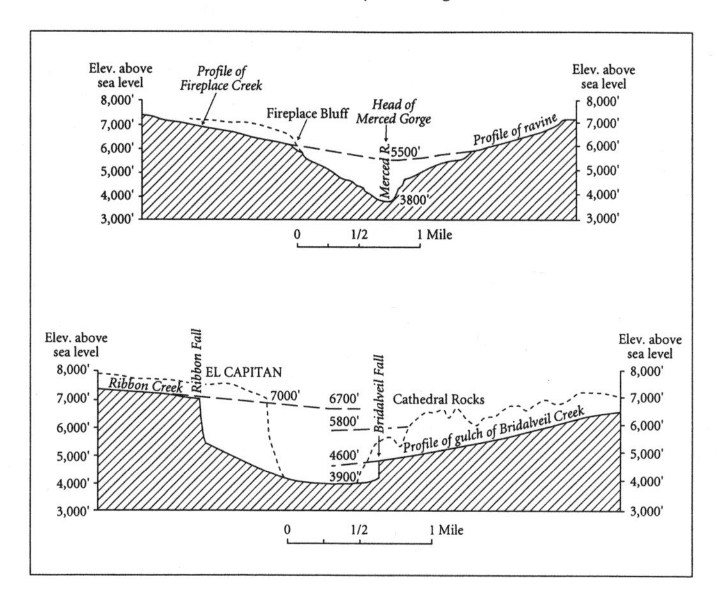

Figure 37. François Matthes re-created the preglacial landscape of the Merced River Canyon by extending to the Merced River gradients of various tributary streams. He attributed differences between the reconstructed valley and the present-day valley to erosion by glaciers. This method is based on some questionable assumptions and is not as valid as Matthes believed. (Drawing by U.S. Geological Survey; Matthes, 1930.)

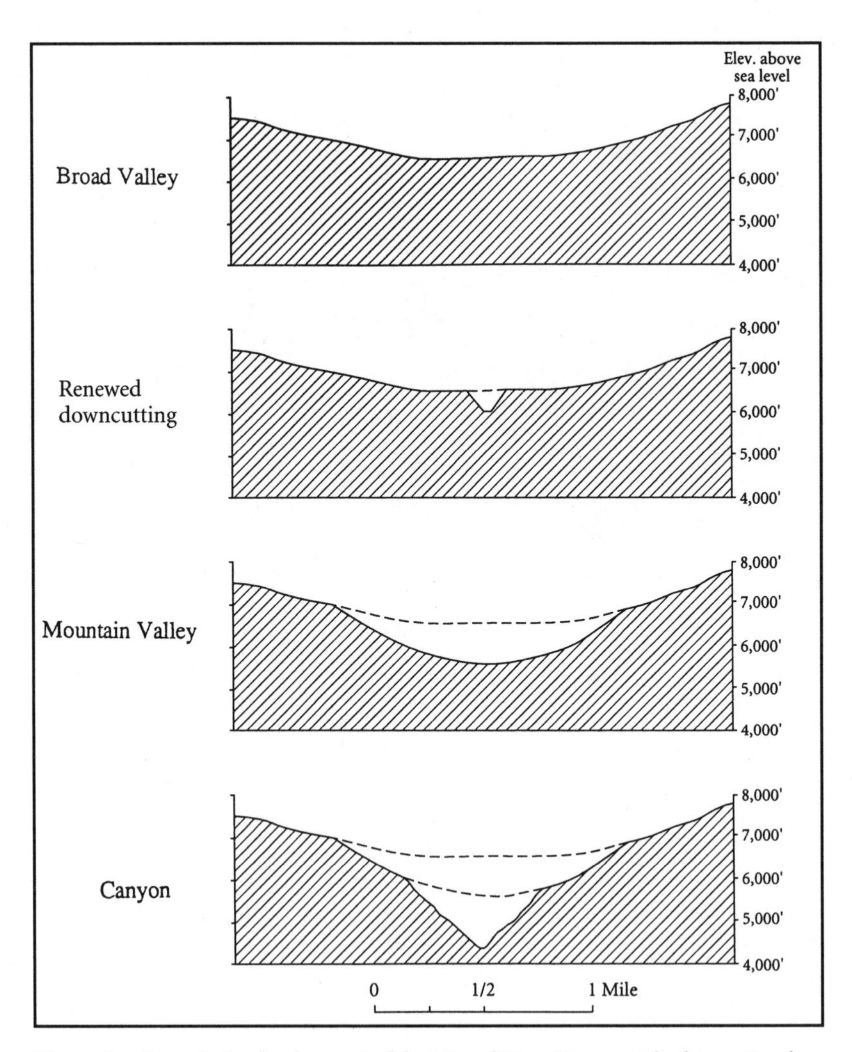

Figure 38. Stages in the development of the Merced River Canyon at the future site of Yosemite Valley as re-created by François Matthes. Today scenes such as these are regarded as single moments in a continually changing scene, not as separate "stages." (Drawing by U.S. Geological Survey; Matthes, 1930.)

THE WIDTH OF YOSEMITE VALLEY

What makes Yosemite Valley special is its width, not its depth, which though striking is not exceptional. Matthes attributed virtually all of the widening of the valley to glacial erosion.

Where a glacier flows through a former river valley it usually widens and straightens the valley. Ice erodes both the sides and the bottom of a valley, whereas a river only erodes the bottom. This is why river valleys are commonly V-shaped in profile while glaciated valleys are commonly U-shaped and have steep valley walls. Also, ice is more viscous than water and does not flow around bends as well, so erosion is concentrated on bends and the valley becomes straighter and more open. The width and the steepness of the cliffs of Yosemite Valley, though especially well developed, are consistent with the character of glaciated river valleys elsewhere.

Dozens of valleys in the Sierra Nevada have been widened by glaciers. John Muir wrote and spoke of "yosemites" as a generic feature and argued that similar valleys were eroded wherever tributary glaciers joined to form trunk glaciers. He wrote, for example, of the "Kings River Yosemite" and the "Tuolumne Yosemite" (Hetch Hetchy Valley). But with all respect to Muir, millions of tourists each year vote with their feet and pocketbooks as they insist that there is only one Yosemite. The widening of Yosemite Valley by ice is a common process done exceptionally well. Why this is so is explained in the next section.

JOINTED ROCK

The importance that jointed rock played in the origin of Yosemite Valley will now be stressed. Studies by Russell, Johnson, and Lawson (see chapter 3) showed that where rock is jointed, glaciers erode effectively, but where joints are lacking, glaciers accomplish little beyond smoothing and polishing. Joints, widened by one or more types of weathering, permit ice to get a "hold" on the rock, so to speak, pull blocks loose, and carry them away; this pulling process is called *plucking* (or *sapping*, or *quarrying*). LeConte and Muir agreed in 1870 that jointing was important in the origin of Yosemite Valley. Matthes included in Professional Paper 160 a section devoted to joints entitled "The Key to the Secret of the Yosemite's Origin." He wrote of the relationship that exists between the sculpture of the cliffs and their "inner structure," pointing out that Yosemite Valley is located

> in an exceptional locality where many small bodies of relatively basic rocks—granodiorite, diorite, and gabbro—have been intruded into the otherwise vast, unbroken bodies of siliceous granite and monzonite that make up the central parts of the great Sierra batholith . . . also that these basic rocks are in general more closely jointed than the siliceous rocks. (Matthes, 1930, p. 91)

Figure 39. Despite the fact that large glaciers flowed through it several times, part of Tenaya Canyon is narrow and V-shaped, very different from Yosemite Valley just a few miles downstream. Lack of joints in the granitic rock prevented the glacier from widening the canyon here.

Explanation is in order. While it is often said that the Sierra Nevada is composed largely of granite, the statement is technically incorrect; here is why. *Granite* is a coarsely crystalline igneous rock with a significant amount of the minerals quartz and potassium feldspar. A similar rock with less of these two minerals and more of the mineral plagioclase is named *granodiorite,* or what Matthes called *monzonite,* even though these rocks still resemble granite in appearance. Rock with essentially no quartz or potassium feldspar is named *diorite* or *gabbro*—the so-called "basic" rocks—in contrast to the "siliceous" rocks, granite and granodiorite. The basic rocks are dark in color and rather different in appearance from the lighter-colored siliceous rocks. As it happens, there is more granodiorite and other siliceous rocks in the Sierra than there is "true" granite. But to write and speak simply, all these light-colored, crystalline rocks are often referred to as granite to distinguish them from the dark-colored, crystalline basic rocks and the multicolored metamorphic rocks. Today it is common to speak and write of "granitic rocks," including those similar in appearance to granite though not necessarily granite in the strict sense.

What do these somewhat technical distinctions have to do with Yosemite Valley? Matthes says that compared to granitic rock, basic rock is more jointed, weathers more rapidly, and so is more easily eroded by glaciers. The local abundance of basic rock enabled the Yosemite Glaciers to be effective.

If basic rock had not been there, the area would consist of "unbroken" granitic rock, glacial erosion would have been less effective, and Yosemite Valley might now resemble nearby Tenaya Canyon (fig. 39) rather than the wide, open, flat-floored valley we regard as special.

Studies of the bedrock in the Yosemite area were done by a colleague of Matthes, F. C. Calkins, who contributed a major section to Professional Paper 160. Huber presents a modern discussion of the bedrock of Yosemite National Park and explains how meanings of some rock names have changed over the years (1987).

LAKE YOSEMITE

Glaciers occupied Yosemite Valley at least three times, and after each retreat a lake probably formed in the valley. Each lake was soon filled with gravel, sand, and mud from the retreating glaciers, the re-forming Merced River, and Tenaya Creek. Matthes named the lake that formed after the most recent glaciation Lake Yosemite.

> To the camper who pitches his tent in one of the Yosemite's shady groves it may not occur, perhaps, that the spot he has selected for his temporary abode once lay in the midst of a mountain lake of exceptional beauty—a sheet of water in which the cliffs of El Capitan and Half Dome as well as the sprays of the Yosemite Falls were reflected. Yet there can be no doubt that the present embowered, parklike floor of the valley replaces such a body of water.
>
> ... [T]he lake owed its existence not to the moraine dam alone; it occupied an elongated basin scooped out in the rock floor of the valley by the ancient Yosemite Glacier, and its depth was merely increased somewhat by the moraine dam which was situated on a broad rock sill at the lower end of the basin. (Matthes, 1930, p. 103)

The last Yosemite Glacier, of Tioga age, was not as large as that of at least one earlier glaciation and did not erode to bedrock. Lake Yosemite was probably smaller and less deep than Matthes believed, though the total sum of river and lake deposits in Yosemite Valley is now known to be greater than he imagined.

DISMANTLING

Matthes identified the final event in the evolution of Yosemite Valley as a period of "dismantling." Dismantling refers to significant changes that the cliffs of Yosemite Valley have experienced since they were last eroded by a glacier (fig. 40). The dominant means of dismantling is rockfall, which can be caused by water freezing in joints, by earthquake shaking, or by the undramatic process of weathering. The Owens Valley earthquake of 1872 produced numerous rockfalls in Yosemite Valley that were witnessed by John Muir.

Figure 40. Royal Arches and Washington Column. Most details of Yosemite cliffs reflect the spacing and shape of joints in the granitic rock and are not unique to erosion by glaciers.

The fatal rockfall from Glacier Point in July 1996 is but the most recent of many such events that have occurred, and that will continue to occur for many millennia. So long as large numbers of people live and camp in Yosemite Valley there is potential for great loss of life and injury from a really big rockfall. Such a threat might justify removal of cabins, stores, and campgrounds from the valley, as was done in Lassen Volcanic National Park when it was recognized that a large campground, cabins, and the visitor center were built on old rock-slide debris of considerable dimension.

Dismantling has created a difference in appearance between Hetch Hetchy Valley and Yosemite Valley. Hetch Hetchy, on the Tuolumne River 18 miles northwest of Yosemite Valley, was one of Muir's beloved "yosemites" before it was dammed and turned into a reservoir in 1923. There was a flat valley floor, waterfall, and steep cliffs, reminiscent of Yosemite Valley. John Muir and the Sierra Club fought hard to prevent the dam, but lost. Some people believe that losing Hetch Hetchy broke Muir's heart and shortened his life.

N. King Huber points out a significant difference between Hetch Hetchy and Yosemite:

> During each major glaciation, including the Tioga, which probably peaked only about 15,000–20,000 yr. ago, the Tuolumne canyon was filled to the brim with ice at least as far west as Mather, some 10 km beyond Hetch Hetchy . . .

Thus, Hetch Hetchy has been glacially scoured "recently." Yosemite Valley, however, has not been filled with ice for at least 750,000 yr., the minimum age of the Sherwin glaciation. . . . Thus, the major excavation of Yosemite Valley, including the bedrock basin beneath the valley floor, had to have been accomplished by that time, and since then, freeze-thaw cycles have promoted spalling of rock slabs, steepening the cliffs, and forming the recessed alcoves into which waterfalls such as Bridalveil now leap. For this reason, Hetch Hetchy Valley is in many ways a "fresher looking" glaciated valley than Yosemite Valley, long considered a classic glacially carved valley.

(Huber, 1990, p. 114)

It should be emphasized that Yosemite Valley has been glaciated since the Sherwin advance but has not been filled with ice since that time. Filled is the key word, for only then could ice shape the walls significantly. The overall shape of Yosemite's cliffs is of glacier origin but their present appearance in detail mostly is not.

STUDIES SINCE MATTHES

François Matthes showed that Yosemite Valley is the combined result of erosion by the Merced River and glaciers of various ages. This conclusion remains the accepted version of the origin of Yosemite Valley, but we have learned things since 1930 that require modification of some of Matthes's work.

Modification number one: In 1935 and 1937 seismic studies were conducted in Yosemite Valley to determine the depth of unconsolidated sediments that overlie bedrock. The results show that the depth to bedrock at the east end of the valley is about 1,800 feet, much greater than Matthes estimated:

The upper layer with a maximum thickness of about 150 m extends from Mirror Lake to the Wisconsin end moraines near Bridalveil Meadow. . . . The intermediate and basal layers have maximum thicknesses of 220 and 300 m respectively, and the intermediate layer lies in a U-shaped trough seemingly gouged out of the basal layer. Both layers are thought to be remnants of earlier lake fillings, and at least the basal layer is pre-Wisconsin. The greatest thickness of fill, about 600 m, is near the head of the valley between Ahwahnee Hotel and Camp Curry.

. . . The amount of glacial excavation on the bedrock floor, essentially double the 450 m previously estimated, is attributed wholly to pre-Wisconsin glaciation. (Gutenberg, Buwalda, and Sharp, 1956, p. 1051)

Modification number two: The old landscapes that Matthes re-created may not have formed as he imagined. As described in chapter 3, Andrew Lawson identified several erosion surfaces in the vicinity of Mount Whitney that he thought marked erosion-to-baselevel during periods of stability that interrupted the uplift of the Sierra Nevada. The old landscapes named by Matthes in the Yosemite area—Broad Valley, Mountain Valley, Canyon—follow this lead. But Clyde Wahrhaftig of the University of California at Berkeley gives reason to suspect that some or all of these old erosion surfaces formed not at a baselevel reached by erosion during a period of stability but behind massive granitic outcrops that do not weather so rapidly as other rock.

> [A] bench, summit flat, or nickpoint can develop in granitic terrane at any altitude at any time. The only requirement is the appearance of solid granitic outcrops to act as local base levels. Once formed, these summit flats are long-persisting and may accumulate ephemeral alluvial deposits at any time after their origin. Other evidence must be used to establish that granitic terranes were once nearly graded plains close to sea level. (Wahrhaftig, 1965, p. 1186)

Modification number three: An assumption of William Morris Davis's that Matthes followed is no longer well regarded by geomorphologists. This is the idea that mountain ranges commonly, or normally, undergo short episodes of uplift followed by long periods of stability during which rivers may erode to baselevel before another uplift begins. In particular, it was not the case with the Sierra Nevada, as is shown by studies by N. King Huber, author of a recent book about the geology of Yosemite National Park (Huber, 1987). This book is written in non-technical language and is beautifully illustrated. The scope of the book is more than glaciers and Yosemite Valley, and it is a good summary of the rocks and geologic history of Yosemite Park. Huber writes about uplift of the Sierra Nevada during the last 25 million years.

> The uplift began slowly and accelerated over time. The range certainly is still rising—and the rate may still be accelerating. The estimated current rate of uplift at Mount Dana, less than 1½ inches per 100 years, may appear small, but it is greater than the overall rate of smoothing off and lowering of the range by erosion. Thus, there is a net increase in elevation. Estimates of uplift amount and rate are based on studies of lava flows and stream deposits thought to be nearly horizontal when formed, but which are now tilted westward toward the Central Valley. Progressive tilt is indicated by older deposits with greater inclinations than younger ones.
>
> François Matthes inferred from his studies that the late Cenozoic uplift occurred in a series of three pulses interrupted by pauses in uplift. In his view each pulse initiated a new cycle of erosion and thus produced a stage of landscape incision characterized by successively greater relief: Matthes' broad valley, mountain valley, and canyon stages. More recent studies show that fortuitous correlation and the commonly local control of erosion weaken Matthes' case for three distinct pulses of uplift. This does not mean that the

uplift was entirely uniform—few things in geology are—but rather that uplift, once initiated, was more nearly continuous than he envisioned.

(Huber, 1987, p. 28)

Modification number four: As described previously, Matthes believed that local abundance of highly jointed basic rock (diorite and gabbro) permitted glaciers to be effective at erosion at Yosemite Valley. Huber points out that most all of the basic rock present in Yosemite Valley is fine grained, rather than coarse grained, and that abundant joints are associated with fine-grained rocks of whatever mineral composition, granitic or basic.

SUMMARY OF THE ORIGIN OF YOSEMITE VALLEY

The geologic story of Yosemite Valley began about 25 million years ago, as slow uplift and westward tilting of the Sierra Nevada enabled the Merced River to erode a valley into the broad western flank of the range. Continued uplift, tilting, and river erosion resulted in an increasingly deep valley, perhaps 2,000 feet below the divides between river valleys. About 1.5 million years ago the Sierra Nevada was high enough, and the climate was suitable, for glaciers to form. At least three times—probably more—glaciers formed along the Sierra crest and flowed along two or more paths into Yosemite Valley to form a trunk glacier. The several Yosemite glaciers widened, straightened, and deepened the river valley into a U-shaped glacial valley with many near-vertical cliffs. Rock with many closely spaced joints was exploited by the glaciers to give the valley its width and depth. Massive rock with few joints endured to form steep, bold cliffs. The most extensive glaciation of which we know (Sherwin) occurred about a million years ago. During this time the maximum depth of glacial erosion was achieved at the east end of Yosemite Valley, and ice completely filled the valley and spilled over the surrounding upland, as at Glacier Point, for example. Widening and deepening of the valley by this largest of Yosemite glaciers produced oversteepened valley walls and hanging valleys where small tributary ice streams could not keep pace. Retreat of this glacier probably left waterfalls and a large, deep, lake that became in-filled with river-transported sediment, burying the lower part of the U-shaped glacial valley. A Yosemite Glacier re-formed at least twice since then but never again became nearly so large as that of a million years ago. The most recent glaciation was at a maximum only about 20,000 years ago, but the glacier was not very large. The ice extended only as far as Bridalveil Meadow and did not erode high on the valley walls. Retreat of this last glacier left moraines; Lake Yosemite; and waterfalls at hanging valleys, such as Yosemite Falls and Bridalveil Fall; and at glacial steps, such as

Vernal Fall and Nevada Fall. Moraines and waterfalls remain, but Lake Yosemite was filled with river-borne sediment to form the nearly flat valley floor on which the Merced River meanders today. During all of this history slow uplift of the Sierra Nevada continued, punctuated by occasional earthquakes that caused rockfalls from the glacially oversteepened walls of Yosemite Valley. Rockfalls continue to fashion cliff details, such as the Royal Arches, as they have done since glacier ice last completely filled the valley, about 1 million years ago. One rockfall created Mirror Lake, now mostly filled in, at the east end of the valley. The rockfall of July 1996 and the flooding of the Merced River in January 1997, so destructive of human "improvements," show that powerful geologic forces remain at large and continue the long evolution of Yosemite Valley.

REFERENCES CITED

Colby, William F. 1950. *John Muir's studies in the Sierra.* San Francisco: Sierra Club.

Gutenberg, B., J. P. Buwalda, and R. P. Sharp. 1956. Seismic exploration on the floor of Yosemite Valley, California. *Geol. Soc. of Amer. Bull.* 67: 1051–78.

Huber, N. King. 1987. *The geologic story of Yosemite National Park.* U.S. Geol. Survey Bull. 1595. (This U.S.G.S. Survey Bulletin is out of print, but a reprint by the Yosemite Association is available in many bookstores.)

———. 1990. The late Cenozoic evolution of the Tuolumne River, central Sierra Nevada, California. *Geol. Soc. of Amer. Bull.* 102: 102–15.

King, Clarence. 1872. *Mountaineering in the Sierra Nevada.* Boston: James R. Osgood and Co.

LeConte, Joseph. [1870; revised 1900] 1971. *A journal of ramblings through the High Sierra of California.* Reprint, New York: Sierra Club/Ballantine Books.

Matthes, François E. 1930. *Geologic history of the Yosemite Valley.* U.S. Geol. Survey Prof. Paper 160.

———. 1938. John Muir and the glacial theory of Yosemite. *Sierra Club Bull.* 23: 9–10.

Muir, John. 1880. Ancient glaciers of the Sierra, California. *Californian* 2: 550–57.

Wahrhaftig, Clyde. 1965. Stepped topography of the southern Sierra Nevada, California. *Geol. Soc. of Amer. Bull.* 76: 1165–90.

Whitney, J. D. 1865. The High Sierra. Chap. 10 in *Report of progress, and synopsis of field work from 1860 to 1864.* Vol. 1 of *Geology.* Philadelphia: Geol. Survey of California.

———. 1868. *The Yosemite book. Geological survey of california.* New York: Julius Bien.

CHAPTER FIVE

ICE AGE GLACIERS OUTSIDE THE SIERRA NEVADA

DISTRIBUTION

Ice Age glaciers existed in five geologic provinces in California other than the Sierra Nevada, as shown in figure 1. Compared to the Sierra Nevada these areas had smaller and less extensive glaciers and have not been studied nearly as much.

> In the northern latitudes of California, glaciers, born in the 9000-foot-high Klamath Mountains moved down their valleys to as low as 2500 feet above sea level. South, in the San Bernardino Mountains, tiny valley glaciers formed in source areas 11,000 feet above the sea and moved downward only to elevations of 8700 feet where melting halted their advance. . . . It is interesting that the Sierra Nevada was extensively glaciated whereas the equally high mountain peaks of Nevada, Arizona, and western Utah contained only small scattered glaciers. The reason for this is that during the Ice Age, as today, these desert regions did not receive enough moisture to allow large ice fields to form even though the climate probably was cold enough to support them.
> (Burnett, 1964, p. 46)

This reminds us that it takes more than cold to form glaciers. One reason that Nevada, Arizona, and Utah are so dry is that California mountains intercept much of the moisture moving east from the Pacific Ocean, creating a rain shadow that extends hundreds of miles eastward. Close to the ocean and with widespread high mountains, California was the most extensively glaciated area in North America south of the great continental glaciers that originated in Canada.

CASCADE RANGE

The Cascade Range, located immediately north of the Sierra Nevada, is characterized by volcanic mountains potentially capable of erupting at any time. Lassen Peak erupted many times between 1914 and 1920. Cinder Cone (in Lassen Volcanic National Park) erupted about 1850, and Mount Shasta erupted in the eighteenth century.

Lassen Volcanic National Park

In September 1863, three months after discovering the first evidence of glaciation in California, William H. Brewer observed evidence of former glaciers in what is now Lassen Volcanic National Park. About 400 square miles around Lassen Park was glaciated (Crandell, 1972), the largest area in California outside of the Sierra Nevada. Glaciers flowed radially from an east-west elongated ice field to lower than 2,400-foot elevation (Hilton and Lydon, 1976). Ice was as much as 1,600 feet thick at Warner Valley in the southeast part of the park (Williams, 1932). Like the Sierra Nevada, the Lassen area has been glaciated more than once. Kane recognizes five glacial episodes and presents a map showing the distribution of deposits formed by the various glacial advances, along with instructive photographs (1982). Gerstel and Clynne find "at least two major episodes of glaciation, each including multiple advances" during the last 75,000 years. The first major episode, named the Reading Peak, had two advances between 58 and 73 thousand years ago; the second major glaciation occurred after Lassen Peak formed between 23 and 28 thousand years ago.

> Four advances are identified within the Lassen Peak episode. During the maximum of the Lassen Peak episode, ice emanating from Lassen Peak extended approximately 8 km into the Hat Creek drainage and 6.5 km into the Lost Creek drainage, to an elevation in both drainages of about 1650 m. Recessional moraines occur in the Lost Creek drainage at 1770 m, 1950 m, and 2380 m. A large early neoglacial moraine characterized by weak soils and a lack of weathering rinds occurs on the SE side of Lassen Peak at 2440 m.
>
> (Gerstel and Clynne, 1989, p. 83)

The "neoglacial moraine" in the quote above refers to glaciation since the end of the Pleistocene, as will be considered more completely in chapter 6. Kelson, Page, Unruh, and Lettis describe how a fault has displaced moraines of three different glacial advances in the area north of Lake Almanor; the estimated ages of these moraines is from 15,000 to 190,000 years (1996).

Mount Shasta

Mount Shasta is another active volcano and a prominent landmark visible throughout northern California. It rises with graceful curves nearly two miles above surrounding lowlands to an elevation of 14,162 feet. About 17 miles in diameter at its base, it has a circumference greater than 50 miles and a volume of about 80 cubic miles, making it by far the largest mountain in California. In contrast, peaks of the Sierra Nevada of slightly higher elevation are but high points on a ridge and lack the massiveness and solitary splendor of Mount Shasta. Being high and large, and having the largest glaciers in California today, one might expect Mount Shasta to have been a major center of Ice Age glaciers, but glaciation around Lassen Park was apparently more extensive even though it is lower in elevation, farther from the coast, and farther south. The explanation is that Shasta is a young volcanic mountain and during much of the Ice Age was lower in elevation than at present; Lassen Peak also is young, younger than Shasta in fact, but the area around Lassen Peak had numerous high mountains during the Ice Age, while there are few high mountains near Mount Shasta.

In 1870 Clarence King found evidence of ancient glaciers up to 12 miles long around Mount Shasta. Quintin Aune describes glacial features in the area:

> The fairly broad valley of the Sacramento River canyon north of Dunsmuir may have been carved in part by the terminal end of a glacier originating from Pleistocene Mt. Shasta, but the evidence is largely buried beneath the lavas of subsequent Mt. Shasta eruptions.
>
> The only clear evidence of former, more intense glacial activity on Mt. Shasta is on the southern, somewhat older portion of the mountain. There, the Mt. Shasta Ski Bowl area and Avalanche Gulch—the latter is the site of rock avalanches and a prominent rock glacier today—are U-shaped valleys, bounded by arête like [*sic*] ridges, evidencing a former period of more extensive glaciation, in latest Pleistocene or early Holocene (Recent) time.
>
> (Aune, 1970b, p. 148)

Robert Christiansen says Mount Shasta consists of four overlapping cones, the oldest of which is over 100,000 years old, and finds evidence of glaciation during several different stages of the volcano's growth (1976).

Medicine Lake Volcano

Thirty miles northeast of Mount Shasta is the Medicine Lake Volcano, also called the Medicine Lake Highlands. It is older than both Lassen and Shasta, has had a more complex volcanic history, and may no longer be active. Anderson describes small cirques, roches moutonnées, striated rock surfaces, and moraines that indicate that the ice was locally 400–500 feet thick in places (1941).

> The greatest accumulation of morainic material is found northwest of Medicine Lake, where the underlying rocks are completely buried and several basins were formed as a result of the irregular deposition. One of the basins is filled with water and is locally called "Little Medicine Lake."
>
> (Anderson, 1941, p. 383)

There is evidence that large floods occurred in the past:

> Catastrophic flooding has eroded a discontinuous network of oversized anastomosing channels on the northwest flank of the Medicine Lake volcano. . . . The flooding was probably triggered by eruption of andesite tuff through a late Pleistocene ice cap on the volcano, about 60,000 to 70,000 or about 130,000 [years] B.P. (Donnelly-Nolan and Nolan, 1986, p. 875)

"B.P." means before present. Flooding caused by eruption of a volcano with glaciers on it is a serious potential hazard at Mount Shasta and other Cascade volcanoes in Oregon and Washington.

KLAMATH MOUNTAINS

The Klamath Mountains of northwest California are geologically similar to the Sierra Nevada but much lower in elevation. The high point, Thompson Peak, is only 8,936 feet in elevation. But being farther north and closer to the ocean than the Sierra Nevada, the Klamath Mountains had glaciers at rather low elevation during the Pleistocene:

> In the northwestern corner of the state, mountains as low as 1,700 m had glaciers on their north sides with cirque floors as low as 1,600 m; farther east in the interior parts of the Klamath Mountains, peaks 2,000–2,100 m had glaciers on their north sides. (Wahrhaftig and Birman, 1965, p. 326)

Areas within the Klamath Mountains where glacial features are present are the Trinity Alps, Castle Crags, and Marble Mountains.

Trinity Alps

The highlands between Thompson Peak and the Trinity River are called the Trinity Alps. The Trinity Alps attract hikers and backpackers because of glacial scenery in an area of about 300 square miles. Moraines, U-shaped valleys, cirques, and glacial lakes plus granitic rock remind many visitors of the High Sierra, but they lie at much lower elevation and cause less shortness of breath (fig. 41).

Figure 41. U-shaped valley, glacial lakes, horns, and arêtes at Stuart Fork Lakes in the Klamath Mountains of northwestern California. (Photo by Charles Scull, U.S. Forest Service.)

> Within the Trinity Alps the major glaciated areas are in the drainages of Coffee, Swift and Canyon creeks, and the Stuart Fork. Less extensive glaciers occupied parts of the drainages of the South Fork of Salmon River, East Fork of Stuart Fork, Rush Creek, and the North Fork of Trinity River. . . .
>
> (Sharp, 1960, p. 308)

> During the Late (Morris Meadow) phase of glaciation, the area mapped contained at least 30 separate valley glaciers and a few cirque glaciers. Three large composite ice streams occupied Swift and Canyon creeks and the Stuart Fork. . . . The largest glacier by area during the Early (Alpine Lake) substage was in Swift Creek. It covered roughly 18 square miles compared to 16.7 and 9.2 square miles respectively for the Stuart Fork and Canyon Creek glaciers.
>
> (Sharp, 1960, p. 328)

The younger glaciers were all less than 10 miles long, but older glaciers appear to have exceeded 13 miles in length. The lowest elevation attained by any of the glaciers was 2,450 feet. Mary Woods contributes two informative articles with annotated photographs of glacial features in Canyon Creek Valley and elsewhere in the Trinity Alps (Woods, 1976 and 1988).

Castle Crags

Castle Crags in the eastern Klamath Mountains, between the Trinity Alps and Mount Shasta, is a group of prominent granite spires that are the central attraction of popular Castle Crags State Park, just off Interstate 5 north of Redding. The Crags and nearby mountains were glaciated but have not been studied in detail. Vennum says Castle Lake is in a cirque, and the north and east sides of the Crags have much glacial scouring and polish, along with several small moraines (1994). Aune describes evidence of two glaciations in the Seven Lakes Basin west of the Crags and states that most peaks over 5,500-feet elevation had cirques (1970a, 1970b):

> Mighty glaciers descended the valleys of Castle and Little Castle Creeks, scouring out the valleys and providing the topographic contrast between the foliated closely jointed serpentine and the massive, resistant granite.
>
> The Crags themselves were capped by numerous small, hanging glaciers which perched atop the newly shaped cliffs and occupied the cirque basins presently occupied by beautiful stands of weeping spruce. Joints in the granite served as master controls for shaping of the granite domes and spires by ice and frost action. (Aune, 1970a, p. 144)

Marble Mountains

The Marble Mountains are between Highway 96 and Scott Valley around Elk Peak (elevation 6,992 feet). Numerous small lakes are of glacial origin. Harms identifies 177 cirques of Pleistocene age in this rather densely vegetated area (1983). The largest cirques were both closest to the coast (40 to 50 miles) and at the lowest elevations.

COAST RANGES

Small glaciers existed in the higher parts of the Coast Ranges about 100–150 miles north of San Francisco Bay. In the vicinity of Snow Mountain,

> [t]hree, and possibly four, isolated peaks in the northern Coast Ranges of California have been glaciated. Altitudes of the peaks range from 6,873 to 7,448 feet and altitudes of cirque floors range from 6,400 to 7,000 feet. Evidence of glaciation consists primarily of rounded bedrock outcrops, striated rock surfaces, and cirque-like forms at the heads of valleys. Total area covered by glaciers probably did not exceed 3 square miles. (Davis, 1958, p. 620)

Glaciers formed on the north-facing slopes of Snow Mountain, Black Butte, and Anthony Peak; Hull Peak may have had a glacier but the evidence is unclear. The glaciers extended to as low as 5,950 feet where a terminal moraine is found on Snow Mountain (Holway, 1911). In the Yolla Bolly–Middle Eel

Wilderness, Mark De Wit found 27 cirques and valleys with erosional and depositional glacial features in an area of 11 square kilometers. Moraines measure as much as 45 feet high and 1,000 feet long, and the lowest elevation of a glacial feature is 5,960 feet (1991).

BASIN AND RANGE PROVINCE

The Basin and Range province is mostly in Nevada, but parts of it extend into California east of the Cascade–Sierra Nevada Ranges. It is a dry region because it is within the rain shadow of more westerly mountains. There are numerous individual mountain ranges formed by faulting that are separated by lowlands. Some of the ranges are quite high, but because of the rain shadow, glaciers were not nearly so extensive as those in the Sierra-Cascade or Klamath Mountains. Three areas that were glaciated are the Warner Mountains, the White-Inyo Mountains, and the Sweetwater Range.

The Warner Mountains, in the little-visited northeast corner of California, is a north-south trending tilted fault block typical of the Basin and Range province. The crest of the block is between 9,000 and 10,000 feet elevation. Despite its northerly location and promising elevation, glaciers were not extensive during the Ice Age because of low snowfall in the rain shadow of the Cascade Range and Klamath Mountains. Duffield and Weldin state that Patterson Lake is a cirque lake, dammed by a moraine, on the northeast side of Warren Peak (9,710 feet elevation) (1976). Hulbe reports moraines as low as 6,400 feet and a glacier 5 miles long in Pine Creek Basin (1980).

The White-Inyo Mountains of east-central California are quite high, with maximum elevation of 14,246 feet at White Mountain and a crest that averages over 12,000 feet in elevation for a distance of 25 miles. But because of the rain shadow of the Sierra Nevada the range is not as extensively glaciated as the high elevation might suggest. However, the White-Inyo Mountains is the most-studied glacial area in California outside the Sierra Nevada. In part this is attributable to the presence of the White Mountain Research Station high in the range south of White Mountain, with a road to it, which offers a good base of operations for researchers. Turner mentions glaciers on the east side of White Mountain leaving distinctive valleys and moraines above 8,000 feet (1900). Blackwelder states that the northern 22 miles of the White-Inyo Mountains show evidence of strong glaciation:

> Small glaciers of the Tioga stage are indicated by fairly clean cirques and suggestions of very bouldery moraines at high elevations. In the Tahoe stage much longer glaciers descended the canyons but none of them approached the base of the range closely. On Perry Aiken Creek, on the east side of the

range, a remnant of what is probably the Tahoe stage moraine stands at an altitude of about 6,500 feet, 2 miles back from the margin of the range.

(Blackwelder, 1934, p. 220)

LaMarche includes a map showing the area covered by former glaciers plus a description of the area:

> Glacially sculptured topography and associated surficial deposits show that several eastward-draining canyons were occupied by small glaciers.... Cirques are located east of the crest from Boundary Peak on the north to Sheep Mountain on the south. Their floors range in altitude from 11,200 feet to 12,500 feet, and many contain tarns and small moraines. Perennial snowbanks and small stratified ice masses cling to the walls of the highest cirques, but there are no glaciers. U-shaped valleys extend eastward from cirques along the crest of the range. Deposits of till as thick as 400 feet make up broad, arcuate steps on the valley floors. On the north fork of Chiatovich Creek there are 3 distinct terminal or recessional moraines which have been deeply incised by axial stream channels; the end moraine farthest downstream is about 4 miles from the cirques. Along the north fork of Cottonwood Creek paired lateral moraines lie 300 feet above the stream; the till, resting on metamorphosed dolomite, contains boulders of granite and metamorphic rocks exposed near the crest of the range. (LaMarche, 1965, p. C144)

Swanson, Elliott-Fisk, Dorn, and Phillips present an interesting idea:

> The White Mountains, CA-NV, were extensively glaciated during the Quaternary, with the extent of glaciation decreasing through time due to tectonic uplift and regional climatic change. We believe that the White Mountains have a similar glacial chronology to the adjacent Sierra Nevada, but that the intensification of the Sierran rain shadow during the Quaternary led to a progressive decrease in glacier size. Thus a more complete glacial record is preserved along the valleys and crestal plateaus of the White Mountains than in the Sierra Nevada, where older glacial deposits have been obliterated by the extensive Tahoe and Tioga glaciers. (Swanson and others, 1988, p. A209)

In the Chiatovich Creek Basin they found evidence of seven glaciations that may be equivalents of the McGee, Sherwin, pre-Tahoe (two episodes), Tahoe, Tioga, and Hilgard glacial episodes in the Sierra Nevada. They are optimistic that the White Mountains may contain the most complete continental record of glacial events in North America. Future study will tell.

The Sweetwater Range is in east-central California, east of Highway 395 and north of Bridgeport. Mount Patterson (11,673 feet) is the point of highest elevation. Blackwelder described the mountains as fairly well glaciated at least during the Sherwin and Tahoe stages. Subdued cirques and small moraines suggesting glacierets of Tahoe age have been found at the heads of Desert and Deep Creeks on the western slope. On the east flank ice tongues of the same age were somewhat longer and more numerous. . . . [T]he glacier in Sweetwater Canyon appears to have been about 3 miles long and extended more than half way down to the base of the range.

(Blackwelder, 1934, p. 220)

SAN BERNARDINO MOUNTAINS

San Gorgonio Mountain (11,502 feet) is the highest mountain in California south of the Sierra Nevada. It is in the San Bernardino Mountains, part of the Transverse Ranges province north of Los Angeles and San Bernardino. Here is the southernmost known-for-certain glaciated area in California, at 34°06′ north latitude. Glacial features were recognized in 1910 (Fairbanks and Carey, 1910) and studied in detail by Sharp, Allen, and Meier (1959). Seven glaciers formed on the north side of a ridge connecting San Gorgonio Mountain with San Bernardino Peak three miles to the west:

> They headed at elevations between 10,300 and 11,300 feet, and none descended below 8700 feet. . . . Lengths were 0.5 to 1.7 miles and areas 0.1 to 0.84 miles. The Dry Lake glacier on the north side of San Gorgonio Mountain was the largest. Steep gradients, as much as 800 to 1000 feet per mile, prevented the ice from attaining a thickness in excess of a few hundred feet.

(Sharp, Allen, and Meier, 1959, p. 84)

POSSIBLE OTHER LOCATIONS

Sharp, Allen, and Meier evaluates reports of glaciation in the San Gabriel Mountains of southern California and argue convincingly that glaciers could not have formed there because of the low elevation (1959). They also considered the possibility of glaciers on San Jacinto Peak near Palm Springs and concluded that the evidence is ambiguous but that this high peak possibly was glaciated.

There probably are reports of glaciated areas in California that I have missed in this compilation, and other areas that have not yet been identified. Not all mountainous areas of California have been studied in detail, and because not all geologists are interested in or knowledgeable about glaciers, even studied areas may contain glacial features as yet unnoticed or unreported.

REFERENCES CITED

Anderson, C. A. 1941. Volcanoes of the Medicine Lake Highland, California. *Calif. Univ. Dept. of Geol. Sci. Bull.* 25: 347–422.

Aune, Quintin A. 1970a. A trip to Castle Crags. *Mineral Info. Service* 23: 139–44.

———. 1970b. Glaciation in Mt. Shasta-Castle Crags. *Mineral Info. Service* 23: 145–48.

Blackwelder, Eliot. 1934. Supplementary notes on Pleistocene glaciation in the Great Basin. *Jour. of the Washington Academy of Sciences* 24: 217–22.

Burnett, John L. 1964. Glacier trails of California. *Mineral Info. Service* 17: 44–51.

Christiansen, Robert L. 1976. Volcanic evolution of Mt. Shasta, California. *Geol. Soc. of Amer. Abstracts with Programs* 8: 360–61.

Crandell, D. R. 1972. *Glaciation near Lassen Peak, northern California.* U.S. Geol. Survey Prof. Paper 800-C, pp. C179–C188.

Davis, S. N. 1958. Glaciated peaks in the northern Coast Ranges. *Amer. Jour. of Science* 256: 620–29.

De Wit, Mark W. 1991. Glacial geology of the Yolla Bolly–Middle Eel Wilderness, northern California. *Geol. Soc. of Amer. Abstracts with Programs* 23, no. 2: 18.

Donnelly-Nolan, Julie M., and K. Michael Nolan. 1986. Catastrophic flooding and eruption of ash flow tuff at Medicine Lake volcano, California. *Geology* 14: 875–78.

Duffield, Wendell A., and Robert D. Weldin. 1976. *Mineral resources of the South Warner Wilderness, Modoc County, California.* U.S. Geol. Survey Bull. 1385-D.

Fairbanks, Harold Wellman, and E. P. Carey. 1910. Glaciation in the San Bernardino Range, California. *Science,* n.s., 31: 32–33.

Gerstel, W. J., and M. A. Clynne. 1989. Glacial stratigraphy in the Lassen Peak area of northern California: Implications for the age of Lassen Peak dacite dome. *Geol. Soc. of Amer. Abstracts with Programs* 21, no. 5: 83.

Harms, Richard William. 1983. Cirques of the Marble Mountains, northwestern California. Ph.D. diss., Univ. of Calif. at Berkeley.

Hilton, R. P., and P. A. Lydon. 1976. Low-elevation glaciation in northern California. *Calif. Geology* 29: 114–16.

Holway, Ruliff S. 1911. An extension of Pleistocene glaciation to the Coast Ranges of California. *Bull. Amer. Geographical Soc.* 43: 161–70.

Hulbe, Christolph W. H. 1980. A note on the geology of the Warner Mountains. In *Geologic guide to the Modoc Plateau and the Warner Mountains: Annual field trip guidebook of the Geological Society of Sacramento,* 149–155. Photocopied.

Kane, Phillip. 1982. Pleistocene glaciation, Lassen Volcanic National Park. *Calif. Geology* 35: 95–105.

Kelson, Keith I., William D. Page, Jeffrey R. Unruh, and William R. Lettis. 1996. Displacement of late Pleistocene glacial deposits by the Almanor Fault near Lassen Peak, northeastern California. *Geol. Soc. of Amer. Abstracts with Programs* 28, no. 5: 80.

LaMarche, V. C., Jr. 1965. *Distribution of Pleistocene glaciers in the White Mountains of California and Nevada.* U.S. Geol. Survey Prof. Paper 525-C, pp. C144–C146.

Sharp, Robert P. 1960. Pleistocene glaciation in the Trinity Alps of northern California. *Amer. Jour. Science* 258: 305–40.

Sharp, Robert P., Clarence R. Allen, and Mark F. Meier. 1959. Pleistocene glaciers on southern California mountains. *Amer. Jour. of Science* 257: 81–94.

Swanson, T. W., D. E. Elliott-Fisk, R. I. Dorn, and F. M. Phillips. 1988. Quaternary glaciation of the Chiatovich Creek basin, White Mountains, CA-NV: A multiple dating approach. *Geol. Soc. of Amer. Abstracts with Programs* 20: A209.

Turner, Henry Ward. 1900. The Pleistocene geology of the south-central Sierra Nevada with especial reference to the origin of Yosemite Valley. Calif. Acad. Sci. Proc., 3rd ser., 1: 261–321.

Vennum, Walt. 1994. Geology of Castle Crags. *Calif. Geology* 47: 31–38.
Wahrhaftig, Clyde, and J. H. Birman. 1965. The Quaternary of the Pacific mountain system in California. In *The Quaternary of the United States,* edited by H. E. Wright and D. G. Frey, 299–340. Princeton, N.J.: Princeton Univ. Press.
Williams, Howel. 1932. Geology of the Lassen Volcanic National Park, California. *Univ. of Calif. Publications in Geological Science* 21: 195–385.
Woods, Mary C. 1976. Pleistocene glaciation in Canyon Creek area, Trinity Alps, California. *Calif. Geology* 29: 109–13.
———. 1988. Ice Age geomorphology in the Klamath Mountains. *Calif. Geology* 41: 273–75.

PART 3

MODERN GLACIERS AND HOLOCENE CLIMATE

CHAPTER SIX

DISCOVERY OF MODERN GLACIERS AND THEIR AGE

NEAR MISSES

On July 2, 1863, William H. Brewer and Charles F. Hoffman climbed and named Mount Lyell in what is now Yosemite National Park. In so doing they walked across the Lyell Glacier, second largest in the Sierra Nevada, but did not recognize it as a glacier.

> Over rocks and snow, the last trees are passed, we get on bravely, and think to be up by eleven o'clock. We cross great slopes all polished like glass by former glaciers. Striking the last great slope of snow, we have only one thousand feet more to climb. In places the snow is soft and we sink two or three feet in it. We toil on for hours. (Brewer, 1966, p. 411)

Just a few days previously Brewer had traced the ghost of the Ice Age Tuolumne Glacier, "the first found on the Pacific slope," but on this day he failed to recognize an actual glacier underfoot. Had he climbed Lyell a month or two later the snow cover might have been diminished enough to expose ice and crevasses, and Brewer might have found in the summer of 1863 the first glacier as well as the first former glacier in California.

In 1866 Clarence King and James Gardner found a glacier near Yosemite National Park, as reported by Josiah D. Whitney they described their findings

> in their notes: "In a deep cul-de-sac which opens southeast on the east slope [of Mount Ritter] lies a bed of ice two hundred yards wide, and about a half a mile long. It has moved down from the upper end of the gorge from thirty to fifty feet this year, leaving a deep gulf between the vertical stone wall and the ice." No such masses of ice were found by the writer or any of his corps at any time, in the higher portions of the Sierra farther south; although such

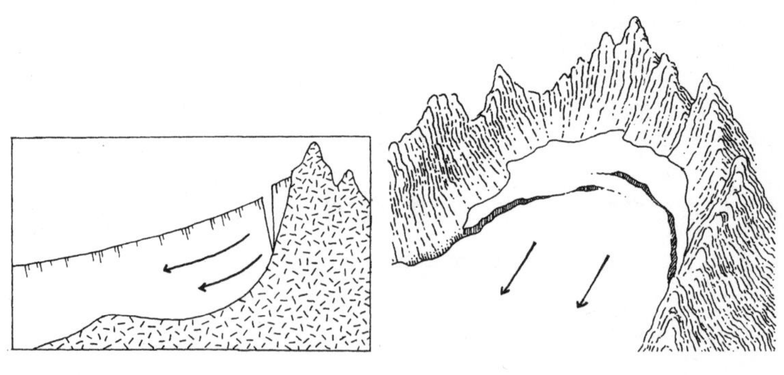

Figure 42. A bergschrund is a crevasse at the head of a mountain glacier where flowing ice pulls away from ice and snow too thin to flow. It is most visible in summer or fall when snow cover is at a minimum. (Drawing by Susan Van Horn.)

> have been reported by others, but not by persons having had any previous experience in the investigation of glacial phenomena. It is doubtful whether these residual masses of ice can with propriety be called glaciers; they have no geological significance as such at the present time, however interesting they may be as possible relics of a once general glaciation of the highest part of the range. (Whitney, 1882, p. 30)

The deep gulf was the bergschrund, a crevasse formed at the head of a mountain glacier where flowing ice pulls away from stagnant ice and the mountain (fig. 42). Bergschrunds help distinguish glaciers, which have definite and significant flow, from perennial masses of ice and snow, which have no significant flow. Had King been older and more independent he might have insisted that he had in fact discovered a glacier in California in 1866, despite Whitney's lack of enthusiasm. The person referred to as inexperienced was surely John Muir, with whom Whitney feuded at length about the importance of glaciers in the Sierra Nevada (see chapter 4).

DISCOVERY ON MOUNT SHASTA

The first official recognition of a glacier in California was on September 11, 1870, by Clarence King, on Mount Shasta:

> [W]e reached the rim of the cone, and looked down into a deep gorge lying between the secondary crater [Shastina] and the main mass of Shasta, and saw directly beneath us a fine glacier, which started almost at the very crest of the main mountain, flowing toward us, and curving around the circular base of our cone. Its entire length in view was not less than three miles, its width opposite our station about four thousand feet, the surface here and there ter-

Figure 43. North side of Mount Shasta. The lower volcanic cone at the far right is named Shastina. The Whitney Glacier extends from near the summit of Shasta diagonally down to the right, between the main mountain and Shastina. (Photo by M. H. Davis, U.S. Forest Service.)

> ribly broken in "cascades" and presenting all the characteristic features of similar glaciers elsewhere. (King, 1871, p. 158)

Most people who had climbed Shasta previously had done so from the south, and no glacier is evident from this route. King was farther to the northwest and saw what he later named the Whitney Glacier (fig. 43). Rhodes (1987) reports that John C. Carroll described a glacier on Mount Shasta in a newspaper article in 1866, but because nothing significant in terms of public knowledge resulted from it, King gets credit for discovery while Carroll gets a footnote.

DISCOVERY IN THE SIERRA NEVADA

John Muir discovered the first glacier in the Sierra Nevada thirteen months after King found his on Mount Shasta:

> On one of the yellow days of October, 1871, when I was among the mountains of the "Merced group," following the footprints of the ancient glaciers that once flowed grandly from their ample fountains, reading what I could of their history as written in moraines, cañons, lakes, and carved rocks, I came

upon a small stream that was carrying mud of a kind I had never seen. In a calm place, where the stream widened, I collected some of this mud, and observed that it was entirely mineral in composition, and fine as flour, like the mud from a fine-grit grindstone. Before I had time to reason, I said, "Glacier mud—mountain meal!"

Then I observed that this muddy stream issued from a bank of fresh quarried stones and dirt, that was sixty or seventy feet in height. This I at once took to be a moraine. In climbing to the top of it, I was struck with the steepness of its slope, and with its raw, unsettled, plantless, new born appearance. The slightest touch started blocks of red and black slate, followed by a rattling train of smaller stones and sand, and a crowd of dry dust of mud, the whole moraine being as free from lichens and weather-stains as if dug from the mountain that very day.

When I had scrambled to the top of the moraine, I saw what seemed to be a huge snow-bank, four or five hundred yards in length, by a half a mile in width. Imbedded in its stained and furrowed surface were stones and dirt like that of which the moraine was built. Dirt-stained lines curved across the snow-bank from side to side, and when I observed that these curved lines coincided with the curved moraine, and that the stones and dirt were most abundant near the bottom of the bank, I shouted "A living glacier!"

(Muir, 1873, p. 69)

Muir called this the "Black Mountain Glacier." His Black Mountain is today named Merced Peak. The mud described is pulverized rock produced by abrasion at the bottom of a glacier. It is very fine and gives streams issuing from glaciers a distinctive milky appearance, and such streams are often referred to as showing "glacier milk."

Muir searched elsewhere:

Then I went to the "snow-banks" of Mts. Lyell and McClure, and, on examination, was convinced they also were true glaciers, and that a dozen other snow-banks seen from the summit of Mt. Lyell, crouching in shadow, were glaciers, living as any in the world, and busily engaged in completing that vast work of mountain-making accomplished by their giant relatives now dead. . . .

On the twenty-first of August last I planted five stakes in the glacier of Mt. McClure. . . . On observing my stakes on the sixth of October, or in forty-six days after being planted, I found that stake No. 1, had been carried down stream eleven inches; No. 2, eighteen inches; No. 3, thirty-four; and No. 4, forty-seven inches. As stake No. 4 was near the middle of the glacier, perhaps it was not far from the point of maximum velocity—forty-seven inches in forty-six days, or one inch per day. Stake No. 5 was planted about midway between the head of the glacier and stake No. 4. Its motion I found to be, in forty-six days, forty inches. Thus these ice-masses are seen to possess true glacial motion.

(Muir, 1873, p. 70)

Mount McClure is today spelled Maclure. It is worth noting that Muir found these glaciers in October, when the previous season's snow had largely melted away. Glaciers are hard to recognize in spring and early summer when most everything is covered with snow.

Muir was not particular about how big a glacier should be; he measured an ice mass on Mount Hoffman that flowed thirteen-sixteenths of an inch in four days:

> This Hoffman glacier is about 1,000 feet long by fifteen to thirty feet wide, and perhaps 100 feet deep in the deepest places. (Muir, 1873, p. 71)

Many persons would be reluctant to regard as a glacier such a small body of ice with such a small amount of apparent motion; the name *glacieret* or *perennial ice mass* is a better choice.

John Muir explored throughout the High Sierra and found many glaciers:

> [E]very cañon and valley was the channel of an ice-stream, all of which may be easily traced back to where their fountains lay in the recesses of the alpine summits, and where some sixty-five of their topmost residual branches still linger beneath cool shadows. (Muir, 1880, p. 551)

It is impossible to know how many of his 65 glaciers were definite, like those on Mounts Lyell and Maclure, and how many were questionable, like that on Mount Hoffman. Sixty-five, or a number close to this, is often cited as the number of glaciers in the Sierra, but the correct number is far from settled (see chapter 7).

THE LITTLE ICE AGE

When glaciers were discovered in the Sierra Nevada it was assumed that they were what was left of the large Ice Age glaciers, but we now know they are only about 700 years old, the product of a climate change within historic times called the Little Ice Age.

Israel C. Russell was the first to consider the age of the Sierra glaciers:

> The present glaciers are perhaps the shrunken remnants of the ancient ice rivers; but, if we follow the teachings of Lake Lahontan, we must conclude that, like the present desert lakes to the east of the Sierra Nevada, they have had a fresh beginning within quite recent times. The Quaternary glaciers of the Sierra may reasonably be supposed to have passed away completely during the arid period which followed the last high water stage of Lake Lahontan. The present glaciers are therefore the result of a modern climatic oscillation, but whether they mark the commencement of a secular period

> of low mean annual temperature or not remains for future observers to decide. (Russell, 1889, p. 326)

Lake Lahontan was a large pluvial lake in northwest Nevada and adjacent California during the Ice Age. Like other lakes in the Great Basin, it is known to have become larger and smaller numerous times during the Pleistocene, as did glaciers in the nearby mountains. Pluvial lakes grow or shrink in response to similar climatic factors that affect mountain glaciers, namely precipitation, temperature, evaporation, and humidity. Russell had no direct evidence of the age of Sierra Nevada glaciers, but knowing that nearby pluvial lakes of Pleistocene age had completely dried up before the present small lakes formed, he reasoned that Sierra glaciers probably had done the same thing.

François Matthes agreed with Russell:

> But recent studies leave little doubt that all remnants of the vast ice mantle which covered the higher parts of the Sierra Nevada during the Ice Age melted away thousands of years ago, and that the present small glaciers represent a new generation born . . . thousands of years after the Ice Age came to an end. (Matthes, 1950, p. 151)

The main reason for believing that this was so is the "Altithermal," also called the "Climatic Optimum," a period of warmer-than-present temperatures that followed the end of the Ice Age in some parts of the world. Maximum average temperatures were reached in North America about 5,000–6,000 years ago, and there is good evidence from many places that the snow line rose a thousand feet or more during this warm period.

> A thousand-foot upward shift of the snow line would not have stripped the ice from all our western mountains. Mount Rainier, the ice-laden summit of which now rises fully 5,000 feet above the snow line, never lost all its glaciers, but the Sierra Nevada surely did, for it has at present no snow line. Even its highest peaks with broad summits bear no perpetual ice. The theoretical snow line lies well above them, probably not less than 2,000 feet above the small glaciers. The latter are able to exist because of the exceptionally favorable conditions which their cirques afford for the entrapment of wind-blown snow in winter and its conservation in summer. Each cirque has a microclimate of its own that renders it just possible for a small ice mass to persist under present conditions. When the snow line rose another thousand feet higher, the timberline zone rose to the cirques, and in the warmth of that zone the last glacierets perished. (Matthes, 1950, p. 157)

Matthes knew that climate in Europe had become much colder a few centuries ago and that during that time glaciers in the Alps had advanced, in some places covering human workings and forcing abandonment of villages that had existed for centuries (Matthes, 1942). Matthes named the cooling that had occurred after the warm Altithermal the "little ice age." He believed the Little Ice Age began about 4,000 years ago, a figure he arrived at by studying Owens Lake in California, just east of the Sierra Nevada. Here is how. The site of Owens Lake is known to have contained a much larger lake during the Pleistocene, and the amount of dissolved solids in the water of Owens Lake is too little to represent the concentration of salts from such a large lake. Analysis indicated that, under existing conditions of inflow and evaporation, it would take about 4,000 years to make such salty water from freshwater of the mountain streams that feed the lake. If Owens Lake was 4,000 years old, then that was when the Little Ice Age began, and that was when the glaciers in the adjacent mountains were born.

Today glacial geologists use the name Little Ice Age in a different sense than Matthes did.

> The term "Little Ice Age" is widely used to describe the period of a few centuries between the Middle Ages and the warm period of the first half of the twentieth century, during which glaciers in many parts of the world expanded and fluctuated about more advanced positions than those they occupied in the centuries before or after this generally cooler interval.
>
> (Grove, 1988, p. 3)

Matthes cited evidence that the climate of Europe was quite cold from the late 1500s to the late 1800s (1939, 1942). Grove has gathered evidence of a worldwide cold interval in the last few centuries, although cooling did not start or end at the same time everywhere (1988). The period A.D. 1250–1900 is largely inclusive of the total cold period.

> It is certainly true that lower temperatures were not sustained throughout the period. The Little Ice Age itself consisted of a series of frequent fluctuations. . . . Such fluctuations consist of individual years and clusters of years for which the weather conditions depart strongly from longer-term means. Average conditions throughout the Little Ice Age were none the less such that mountain glaciers advanced to more forward positions than those they had occupied for several centuries, or in some areas even millennia, and fluctuated about those positions until the warming phase in the decades around the turn of the century brought them back to where they had been in earlier Holocene warm periods. (Grove, 1988, p. 5)

To recapitulate, Matthes thought there had been a renewed glacial episode during the last few thousand years since the warm spell called the Altithermal. He called this the Little Ice Age. Modern usage regards the Little Ice Age as cooling during only the last seven centuries. How then do we refer to glacial advances that may have occurred after the Altithermal but before the Little Ice Age? Answer: neoglaciation (new glaciation).

NEOGLACIATION

Table 4 shows names in use for different parts of the Holocene Epoch, the 10,000 years since the end of the Pleistocene Epoch. The early Holocene was a transitional period. The Altithermal was a warm time when glaciers in California totally disappeared. Neoglaciation refers to a time of cooling when glaciers advanced or formed in many parts of the world. The Little Ice Age is the most recent of several cooling periods during neoglaciation.

TABLE 4. Subdivisions of the Holocene Epoch, the last 10,000 years since the end of the Pleistocene Epoch.

Holocene Epoch	Modern time (since 1900)	Most glaciers are retreating
	Neoglacial	Little Ice Age (1250–1900)
		Glaciers advanced and retreated several times at different places
	Altithermal	A warm period
	early Holocene	
Pleistocene Epoch		

HOLOCENE GLACIERS OF CALIFORNIA

With the foregoing as background we can focus on California. Current thinking is that existing glaciers in California were born during the Little Ice Age and are only about 700 years old. In honor of François Matthes, many glacial geologists refer to these glaciers as Matthes glaciers, the product of the Matthes glacial advance that occurred during the Little Ice Age. During the Little Ice Age, temperatures fluctuated and California glaciers advanced and retreated in response. The most recent glacial maximum of the Little Ice Age occurred between 1895 and 1897 (Curry, 1969). This was close to the time when King and Muir were discovering glaciers in California.

For many years geologists thought there were three glacial advances since the end of the Pleistocene, named the Hilgard, the Recess Peak, and the Matthes (Birman, 1964). But in recent years doubt has been cast on the validity of the idea that the Hilgard and Recess Peak belong to the Holocene. For the Hilgard, new mapping and radiocarbon dates show it is a part of the Tioga glaciation of the Pleistocene (Clark and Clark, 1995). The Recess Peak advance was questioned by Burbank (1991), and it has been dated as having ended about 13,000 years ago (Clark, 1995; Clark and Gillespie, 1994):

> Recess Peak therefore is late Pleistocene and predates the European Younger Dryas event. The absence of any glacial deposits between the Recess Peak and Matthes deposits in the Sierra demonstrates that: (1) the Younger Dryas was not a glacial event in central California; (2) the Matthes advance was the most extensive, and possibly the only, Neoglacial event in the range; and (3) climate in the Sierra between [about] 13,000 and 700 yr. ago was too warm and/or dry to support significant glaciers. (Clark and Gillespie, 1994, p. A447)

The second item in the quote above implies that if there were Neoglacial glaciers before the Matthes, they were small, and that the larger glaciers of the Matthes advance pushed aside and destroyed their moraines, leaving no direct evidence of their existence.

REFERENCES CITED

Birman, Joseph H. 1964. *Glacial geology across the crest of the Sierra Nevada, California.* Geol. Soc. of Amer. Special Paper 75.

Brewer, William H. 1966. *Up and down California in 1860–1864.* Edited by F. P. Farquhar. 3rd ed. Berkeley: Univ. of Calif. Press.

Burbank, Douglas W. 1991. Late Quaternary snowline reconstructions for the southern and central Sierra Nevada, California, and a reassessment of the "Recess Peak glaciation." *Quaternary Research* 36: 294–306.

Clark, Douglas H. 1995. Extent, timing, and climatic significance of latest Pleistocene and Holocene glaciation in the Sierra Nevada. Ph.D. diss., Univ. of Washington, Seattle.

Clark, Douglas H., and Gillespie, Alan R. 1994. A new interpretation for late-glacial and Holocene glaciation in the Sierra Nevada, California, and its implications for regional paleoclimate reconstructions. *Geol. Soc. of Amer. Abstracts with Programs* 26: A447.

Clark, Douglas H., and Malcolm M. Clark. 1995. New evidence of late-Wisconsin deglaciation in the Sierra Nevada, California, refutes the Hilgard glaciation. *Geol. Soc. of Amer. Abstracts with Programs* 27, no. 5: 10.

Curry, Robert R. 1969. Holocene climatic and glacial history of the central Sierra Nevada. In *United States contributions to Quaternary research,* edited by S. A. Schumm and W. C. Bradley. Geol. Soc. of Amer. Special Paper 123: 1–47.

Grove, Jean M. 1988. *The Little Ice Age.* London: Methuen.

King, Clarence. 1871. On the discovery of actual glaciers on the mountains of the Pacific slope. *Amer. Jour. of Science and Arts,* 3rd ser., 1: 157–67.

Matthes, F. E. 1939. Report of Committee on Glaciers, April, 1939. *Transactions of the American Geophysical Union* 20: 518–23.

———. 1942. Glaciers. In *Physics of the earth,* edited by O. E. Meinzer. Part 9, Hydrology, 149–219. New York: McGraw-Hill.

———. 1950. *The incomparable valley: A geological interpretation of the Yosemite.* Edited by F. Fryxell. Berkeley: Univ. of Calif. Press.

Muir, John. 1873. On actual glaciers in California. *Amer. Jour. of Science and Arts* 5, no. 105: 69–71.

———. 1880. Ancient glaciers of the Sierra, California. *Californian* 2: 550–57.

Rhodes, Philip T. 1987. Historic glacier fluctuations at Mount Shasta, Siskiyou County. *Calif. Geology* 40: 205–11.

Russell, Israel C. 1889. *Quaternary history of Mono Valley, California.* U.S. Geol. Survey 8th Annual Report, 261–394.

Whitney, J. D. 1882. The climatic changes of later geological times. *Contributions to American Geology.* Vol. 2. Cambridge, Mass.: Museum of Comparative Zoology.

CHAPTER SEVEN

MODERN GLACIERS

DISTRIBUTION

Glaciers exist today in California on Mount Shasta, in the Trinity Alps of the Klamath Mountains, and in thirteen different areas of the Sierra Nevada (fig. 44). Because of its great height and northerly location, Mount Shasta rises above the permanent snow line and has glaciers or glacierets on all sides. Other California glaciers are below the permanent snow line and can survive the warm and sunny California summers only in north-facing cirques, where snow accumulation is enhanced and ablation is retarded (fig. 45).

> [S]uch cirque glaciers owe their existence primarily to local topographic influences on snow accumulation, rather than to regional climate trends. Local excess accumulation is most pronounced in deep, north-facing cirques located next to, and typically downwind of, major divides, especially divides downwind of large canyons that act as storm conduits. These sites are commonly subject to intense snow avalanching or drifting during the accumulation season, and extensive shading by headwall cliffs during the summer.
>
> (Clark, Gillespie, and Clark, 1993, p. A156)

Glaciers that owe their existence to excess accumulation of snow blown about by wind are sometimes referred to as drift glaciers.

François Matthes describes the glaciers of the Sierra Nevada:

> The longest of them measure less than a mile in length, but they are nevertheless glaciers in the true sense of the term and are not mere snow fields, for they are composed of hard, granular ice, they move slowly downhill with a slow flow-like motion, and they are broken by crevasses in consequence of their motion over irregular beds. Moreover, just as is the case with . . . large glaciers . . . so at the extreme head of each of these small glaciers there is

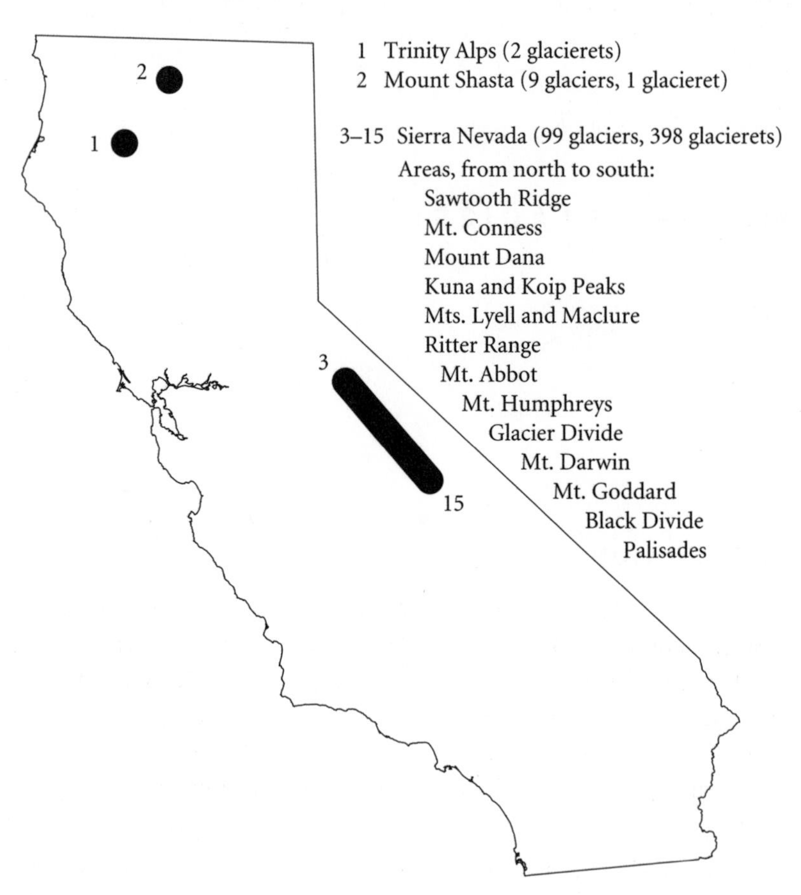

Figure 44. Location of existing glaciers and glacierets in California. The total is 108 glaciers and 401 glacierets. (After Burnett, 1964, and Hill, 1975.)

Figure 45. Glaciers in eastern Yosemite National Park. At front left are Glacier Canyon, Mount Dana, and Dana Glacier. In the right distance is the Lyell Glacier (east and west parts), and to the right of it is the Maclure Glacier (partly in shadow). Glaciers in the left distance are near Koip Peak. During the Ice Age an ice field covered the area and only the highest peaks protruded above the ice. (Photo by M. F. Meier, U.S. Geological Survey.)

> between the ice and the rock of the mountain a gaping cleft (the *bergschrund* of Swiss mountaineers) that opens periodically as a result of the forward movement of the ice. (Matthes, 1962, p. 125)

HOW MANY GLACIERS ARE THERE IN CALIFORNIA?

The short answer is: 108 glaciers and 401 glacierets. But the long answer is more informative and explains why markedly different answers to the question are given by different sources, as well as why no short answer is adequate.

The long answer is: no one knows, or at least no one agrees. Here are some estimates. John Muir found "about" 65 in the Sierra Nevada; Clarence King reported 5 on Mount Shasta; Robert Sharp found 2 in the Trinity Alps, therefore the total is "about" 72. Rhodes, however, counted 10 on Mount Shasta (1987), and Hershey counted 3 in the Trinity Alps (1903). Some of Muir's 65 glaciers are known to have melted. But Muir surely did not find every glacier in the Sierra. Hill states that California has 80 glaciers (1975). Raub, Post, Brown, and Meier state there are 497 glaciers in the Sierra

Nevada alone (1980). From the foregoing it appears that California has somewhere between 72 and 510 glaciers.

Counting glaciers obviously is not a simple task. There are several reasons for the difficulty. Despite dictionaries and glossaries, there is no compulsion for all investigators to use the same definition as to exactly what a glacier is. It is sometimes hard to tell a small glacier from a perennial mass of snow and ice. Small glaciers and perennial ice masses can change one into the other as the amount of snowfall varies from year to year. Finally, many ice bodies in the Sierra Nevada are difficult to get to for direct observation.

The most recent and thorough glacier count for the Sierra Nevada is that of Raub, Post, Brown, and Meier. They used aerial photographs in an effort to distinguish glaciers from long-lived snowfields:

> "Glaciers" were defined as ice masses having a bergschrund or crevasses; visible snow, firn or ice of different years; and/or moraines; 497 were measured, ranging from less than 0.1 km^2 to 1.58 km^2 in area. In addition, about 847 small perennial "ice patches" were classified.
>
> (Raub, Post, Brown, and Meier, 1980, p. 33)

Of the 497 glaciers, the lowest is at 2,769 meters (9,082 feet), the highest at 4,267 meters (13,996 feet), and the mean elevation is 3,543 meters (11,621 feet). Some people are uneasy with these results. The smaller and lower ice masses are . . . well, they are awfully small and awfully low. The presence of a moraine is inconclusive; it may be relict, not formed by the existing ice, or it may be a fake, like a protalus rampart where marginal morainelike debris accumulates by rocks falling and sliding into position instead of being transported and deposited by flowing ice. Conspicuous by absence from their definition is the traditional idea that a glacier consists of flowing ice. Results of their work were originally presented before a live audience and were later published along with remarks by members of the audience. One attendee asked why, even in an aerial-photo study, flow was not considered, as evidenced by mud in meltwater. The answer was that this was attempted but was difficult to apply. Another person in the audience was distressed by their definition of a glacier, noting that if it were used in Japan there would suddenly be glaciers where there were none before, a circumstance "too sensational for mass communication." We might have a media-driven Ice Age upon us!

Back to the question "How many glaciers are there in California?" Surely we can do better than saying "Between 72 and 510!"

The following four-step line of reasoning yields a definite answer, although it is somewhat arbitrary. (1) I distinguish between a glacier and a glacieret. A glacieret is regarded as either a very small glacier or a tiny mass of ice or firn in high mountains resembling a glacier but defying a precise definition. (2) For the Sierra Nevada, I consult U.S.G.S. 15′ topographic quadrangles and count as glaciers the number of glacier symbols on them that have at least one dimension measuring a quarter mile or more. (3) I respect the objectivity and thoroughness of the work of Raub and his associates; their count of 497 candidates is an important contribution even though I am reluctant to accept all of them as glaciers; accordingly I subtract the number of glaciers obtained in step 2 above from 497 to get the number of glacierets in the Sierra Nevada. (4) Rhodes convinces me there are more than 5 glaciers on Mount Shasta, but I think 1 of his 10 is a glacieret. I count the 2 ice bodies in the Klamath Mountains as glacierets.

Here is the result:

Sierra Nevada	99 glaciers	398 glacierets
Mount Shasta	9 glaciers	1 glacieret
Klamath Mountains	0 glaciers	2 glacierets
California Total	108 glaciers	401 glacierets

I believe that saying there are 108 glaciers and 401 glacierets in California is objective, fair, and meaningful.

SELECTED GLACIERS

Some California glaciers are better known than others, in part because of accessibility. Reaching some glaciers or glacierets in the Sierra Nevada requires considerable mountaineering skills, and these are rarely visited; possibly some have never been trod upon. Others, for example the Lyell Glacier in Yosemite National Park, are visited by many people every year and have been studied regularly since 1871.

GLACIERS OF MOUNT SHASTA

Mount Shasta, 14,162 feet, is an active volcano about 50 miles north of Redding whose summit, covered with ice and snow, is visible throughout northern California (fig. 46). It is the only mountain in California that rises above the permanent snow line, the elevation above which snow does not melt during a summer even on exposed slopes. It has the longest glacier in California (Whitney Glacier) and the one with the most volume of ice (Hotlum), plus 7 other glaciers and 1 glacieret. Impressive icefalls are present on the Whitney, Hotlum, and Wintun Glaciers (Selters and Zanger, 1989). Mount Shasta has 74 million square feet of ice and snow, with a volume of 4.7 billion cubic feet (Driedger and Kennard, 1986).

The glaciers on Mount Shasta are not very old.

Figure 46. Mount Shasta viewed from the northwest, showing Whitney (W), Bolam (B), and Hotlum (H) Glaciers, August 1981. (Photo by C. L. Driedger, U.S. Geological Survey.)

> The Hotlum cone, forming Mt. Shasta's summit and undissected north and northeast flanks, postdates early neoglacial deposits and is overlain by no glacial deposits older than a few centuries; weak soil oxidation and lack of cirques also indicate an age of less than a few thousand years.
>
> (Christiansen, 1976, p. 360)

Clarence King, with Fred Clark, explored the McCloud (Konwakiton) Glacier in 1871 and got the scare of his adventurous life. They climbed onto the moraine at the terminus of the glacier:

> In irregular curve it continues not less than three miles around the end of the glacier, and in no place that I saw was less than a half mile in width. Where we had attacked it the width cannot be less than a mile, and the portion over which we had climbed must reach a thickness of five or six hundred feet. . . .
>
> Before long we came to a region of circular, funnel-shaped craters, where evidently the underlying glacier had melted out and a whole freight of boulders fallen in with a rush. Around the edges of these horrible traps we threaded our way with extreme caution; now and then a boulder, dislodging under our feet, rolled down into these pits, and many tons would settle out of sight. Altogether it was the most dangerous kind of climbing I have ever seen.
>
> (King, 1872, pp. 263–64)

Shortly King and Clark were on the glacier proper, where, after some difficulties, they stopped for lunch.

So when we got into a pleasant, open spot where the glacier became for a little way smooth and level we sat down, leisurely enjoying our repast. We saw a possible way out of our difficulty, and sat some time chatting pleasantly. When there was no more lunch we started again, and only three steps away came upon a narrow crack edged by sharp ice-jaws. There was something noticeable in the hollow, bottomless darkness seen through it which arrested us, and when we had jumped across to the other side, both knelt and looked into its depths. We saw a large domed grotto walled in with shattered ice and arched over by a roof of frozen snow so thin the light came through quite easily. The middle of this dome overhung a terrible abyss. A block of ice thrown in fell from ledge to ledge, echoing back its stroke fainter and fainter. We had unconsciously sat for twenty minutes lunching and laughing on the thin roof, with only a few inches of frozen snow to hold us up over that still, deep grave. A noonday sun rapidly melting its surface, the warmth of our persons slowly thawing it, and both of us playfully drumming the frail crest with our tripod legs. We looked at one another, and agreed that we lost confidence in glaciers.

(King, 1872, p. 267)

The cavern may have been a crevasse or a cavern melted by a steam vent or a hot spring, which are common on active volcanoes. Persons visiting glaciers should be roped together and should dine at separate tables.

Clarence King recognized 5 glaciers on Mount Shasta, and this number is often repeated, but Philip Rhodes identifies 10 glaciers (fig. 47) (1987). King's 5 are the Whitney, Hotlum, Bolam, Wintun, and Konwakiton (McCloud). New glaciers of Rhodes are the Stuhl (or Mud Creek), Chicago, Upper Wintun, Watkins, and Olberman. Rhodes describes each of the 10 glaciers and justifies recognition of the new 5, though he states the Olberman Glacier is active only during times of above-average snow accumulation.

Whitney glacier: Largest and longest glacier in California. Only well defined valley glacier in California. Named in 1870 by Clarence King after then State Geologist J. D. Whitney. King claimed to have discovered Whitney glacier in 1870 but an early day climber described it in an 1866 newspaper article (Carroll, 1866). Since circa 1940, Whitney glacier has undergone rejuvenation similar to that observed at Nisqually glacier on Mount Rainier.

(Rhodes, 1987, p. 207)

The Whitney Glacier is 2 miles long, is quite steep, and has ice 126 feet thick (Driedger and Kennard, 1986).

Russell presents a map and several drawings of Shasta's glaciers (1885). Diller says the Whitney Glacier was 2½ miles long and extended down to an elevation of 9,500 feet (1895); like glaciers almost everywhere, it is smaller today. Aerial photos taken of Mount Shasta in 1920 and in 1935 show that over half the ice on the mountain vanished within that fifteen-year interval

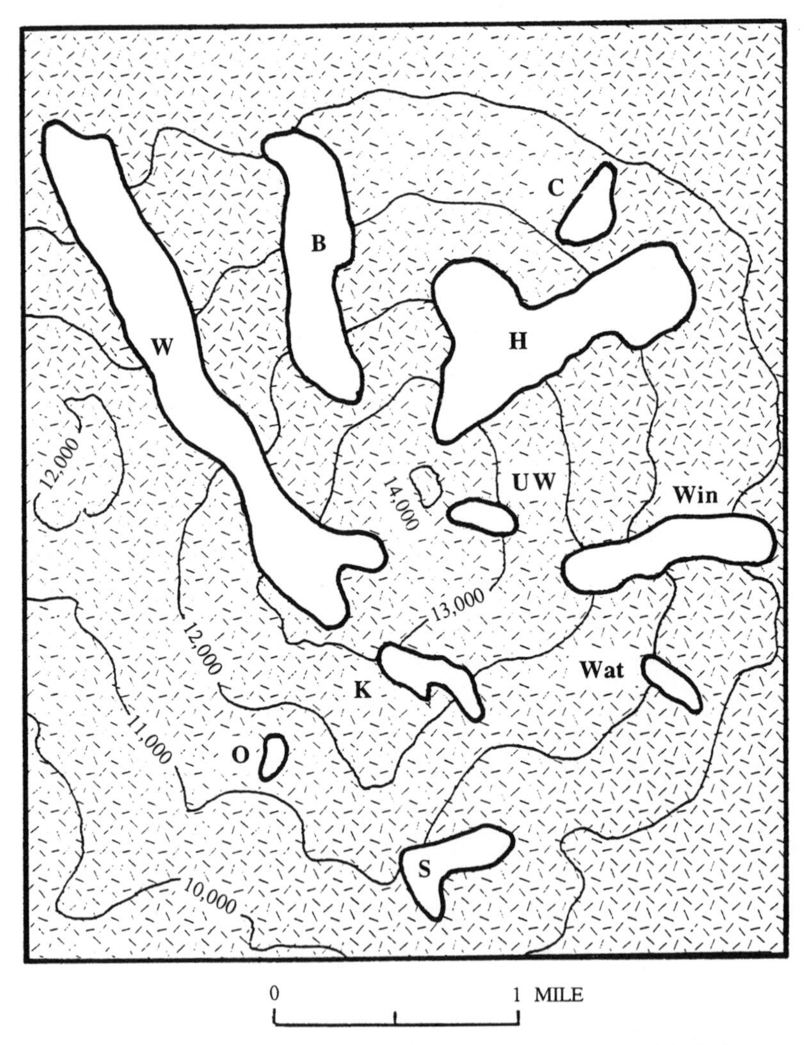

Figure 47. Map of the glaciers of Mount Shasta: Whitney (W), Bolam (B), Hotlum (H), Chicago (C), Upper Wintun (UW), Wintun (Win), Watkins (Wat), Konwakiton (K), Stuhl (S), Olberman (O). Contour lines are in feet. (After Rhodes, 1987.)

(Matthes, 1942). But Selters and Zanger report that the Whitney Glacier became 1,500 feet longer between 1944 and 1972 (1989). Mary Hill and Ernest Carter present photographic accounts of climbing the Whitney Glacier that show moraines, an icefall, seracs, a bergschrund, and an aerial view of the entire glacier (Hill, 1978; Carter, 1984).

In the summer of 1924 mudflows moved down Mud Creek on the southeast slope of Mount Shasta into the McCloud River (Hill and Egenhoff, 1976). The mud covered an area as much as 5 miles long and a mile wide to

Figure 48. Ash Creek and the terminus of the Wintun Glacier, Mount Shasta. A volcanic eruption could melt the glacier and create a large mudflow that might extend for miles from the mountain. (Photo by C. D. Miller, U.S. Geological Survey.)

a depth of 10 feet, closed roads in the area, and created serious problems for the water supply of the town of McCloud. Investigation revealed the mudflow was caused by the breakup and melting of the Konwakiton Glacier (Rhodes says it was the Stuhl Glacier [1987]). Similar outburst floods (jökulhlaups) occurred again in 1926 and in 1931. Hill and Egenhoff report that outburst floods have occurred often in the past, dating from about A.D. 800 and as recently as 1881 and 1920.

Mount Shasta is an active volcano that has erupted on average every 600 years during the last 4,500 years (Miller, 1980); the last eruption was 200 years ago. Glaciers tend to occupy depressions into which lava would be funneled during an eruption (fig. 48), and they constitute a geologic hazard because of rapid melting that might accompany an eruption, resulting in possibly destructive flooding or mudflows. If all the ice and snow on Mount

Shasta melted during a major eruption, approximately 100,000 acre-feet of water would be produced, enough to cause major flooding in the lowlands. In the past, flooding has produced mudflows that extended as far as 16 miles from the summit of Mount Shasta (Driedger and Kennard, 1986). A future volcanic eruption with glacier meltdown is a hazard for several cities in northern California. Because such events are rare it is often difficult for us to take the threat seriously, but major flooding was produced by melting glaciers when Mount St. Helens erupted in 1980, and such flooding can occur in California.

PALISADE GLACIER

The Palisade Glacier (figs. 49–54) is located 14 miles southwest of Big Pine in a cirque on the northeast flank of North Palisade, a 14,242-foot peak on the crest of the Sierra Nevada. With a surface area of about one-half square mile, it is the largest glacier in the Sierra Nevada. Published studies are by Von Engeln (1933), Harrison (1960), and Trent (1983). Von Engeln describes the glacier and presents numerous photographs showing various features of the glacier. Harrison presents a photograph of the glacier that is of historic value.

D. D. Trent writes,

> It lies at elevations between 12,000 feet (3,700 m) and 13,400 feet (4,100 m) in the shadow of impressive 14,000 foot (4,300 m) peaks and covers an area of barely one-half square mile (1.3 km square). Visiting the Palisade glacier is, in a sense, like stepping back in time to get a glimpse of the Sierra Nevada during the great Ice Age of the Pleistocene Epoch. The scene is one of strictly ice, snow, and rock; no vegetation, not even lichen can be found adjacent to the glacier.
>
> The glacier, although small by some standards, exhibits all of the well-known glacial features: crevasses, moraines, a load of rock debris, and a bergschrund. . . . Late in the melting season the snow on the lower part of the glacier may be completely melted to expose hard glacial ice laden with dirt and rock debris. (Trent, 1983, p. 264)

He describes various features of the Palisade Glacier, including glacier tables, crevasses, and a *moulin.* A moulin is a vertical hole in the ice with a stream of meltwater flowing into it; the water flows through or under the ice and reappears at the terminus of the glacier. Moulins are dangerous and should be approached only with great care. Glacier tables form when a flat rock measuring a foot or more across falls onto a glacier and protects the snow or ice beneath it from solar radiation while the surrounding snow or ice melts (or sublimates), leaving the rock on a pedestal. Glacier tables have been found in the Sierra with rocks up to 34 feet across and 10

Figure 49. Palisade Glacier, largest in the Sierra Nevada, in 1977. The crevasse in shadow close to the mountain is the bergschrund. Bare ice is prominent below the firn line. (Photo by Walter Stephens.)

feet thick, with ice pedestals 8 feet high and 8 feet in diameter. Small pieces of rock a few inches across do not form tables because they absorb solar radiation and conduct heat to underlying snow or ice, melt it, and sink into a hole.

Trent and his associates have measured the rate of flow of the Palisade Glacier as a maximum of 23 feet per year. This is a little less than an inch a day, about the same as that which John Muir measured for the Maclure Glacier in the early 1870s. Trent includes both an aerial photo and a map of the glacier, as well as two photographs taken from the same location, which record major shrinkage of the glacier between 1940 and 1980.

(text continues on page 127)

Figure 50. Meltwater stream on the surface of the Palisade Glacier. (Photo by Walter Stephens.)

Figure 51. Meltwater streams often flow into moulins like this one on the Palisade Glacier. (Photo by Walter Stephens.)

Figure 52. Matthes (Little Ice Age) moraines on the Palisade Glacier. (Photo by Walter Stephens.)

Figure 53. First ascent of a glacier table, Palisade Glacier. (Photo by Walter Stephens.)

Figure 54. Surveying on the Palisade Glacier. Making good maps and measuring rate of flow require surveying gear, which is heavy and bulky. (Photo by Walter Stephens.)

Figure 55. Middle Palisade Glacier, Sierra Nevada. (Photo by George Hafer.)

MIDDLE PALISADE GLACIER

Three miles southeast of Palisade Glacier is the Middle Palisade Glacier (fig. 55). At 37°04′ northern latitude, it is the southernmost named glacier in the Sierra Nevada. Several glacier symbols are shown south of the Middle Palisade Glacier on U.S.G.S. 15′ topographic quadrangles, but these are glacierets.

BLACK MOUNTAIN GLACIER

This former glacier, the first recognized in the Sierra Nevada, was found and named by John Muir in 1871. Pierce and Pierce give hiking instructions to visit the site on the north side of Merced Peak (Muir's Black Mountain) in southern Yosemite National Park (1973). Seventy-eight years after Muir discovered the glacier,

> glacial milk in the lakelet below the cirque in 1949 prompted Alfred R. Dole and Richard M. Leonard to explore the glacier. Ice was still present in good quantity, but they felt the glacier, one of the lowest in the Sierra, should probably be classed as "fossil" or inactive. (Leonard, 1951, p. 130)

The elevation of the site is only 11,000 feet; most Sierra glaciers are over a thousand feet higher. Jeffrey Schaffer reports that

> [b]y late summer 1977, after two dry years, the Merced Peak snowfield had completely disappeared. (Schaffer, 1992, p. 193)

This change from being a prominent glacier in 1871 to being a glacieret in 1949 and to being nonexistent by 1977 is in keeping with a worldwide warming trend since about the year 1900.

GLACIERS OF THE RITTER RANGE

Ten glaciers and several glacierets are present on the Ritter Range southeast of Yosemite National Park (figs. 56, 57). The most prominent peaks are Mount Ritter (13,157 feet), Banner Peak (12,945 feet), and the Minarets (more than 12,000 feet). The longest glacier, about three-quarters of a mile, flows northwest from Mount Ritter. This glacier is described by John Muir in recounting his ascent of Mount Ritter in 1872:

> Arriving on the summit of this dividing crest, one of the most exciting pieces of pure wilderness was disclosed that I ever discovered in all my mountaineering. There, immediately in front, loomed the majestic mass of Mount Ritter, with a glacier swooping down its face nearly to my feet, then curving westward and pouring its frozen flood into a dark blue lake, whose shores were bound with precipices of crystalline snow; . . . I could see only the one sublime mountain, the one glacier, the one lake; the whole veiled with one blue shadow—rock, ice, and water close together without a single leaf or sign of life. After gazing spellbound, I began instinctively to scrutinize every notch and gorge and weathered buttress of the mountain, with reference to making the ascent. . . . The head of the glacier sends up a few finger-like branches through narrow *couloirs;* but these seemed too steep and short to be available, especially as I had no ax with which to cut steps, and the numerous narrow-throated gullies down which stones and snow are avalanched seemed hopelessly steep, besides being interrupted by vertical cliffs; while the whole front was rendered still more terribly forbidding by the chill shadow and the gloomy blackness of the rocks.
>
> . . . I . . . climbed out upon the glacier. There were no meadows now to cheer with their brave colors, nor could I hear the dun-headed sparrows, whose cheery notes so often relieve the silence of our highest mountains. The only sounds were the gurgling of small rills down in the veins and crevasses of the glacier, and now and then the rattling report of falling stones, with the echoes they shot out into the crisp air. (Muir, 1894, p. 49)

Muir reached the summit of Mount Ritter only after nearly falling to his death. He describes his descent of a glacier on the south side of the mountain:

Figure 56. Glacier on the Minarets, Ritter Range, Sierra Nevada, viewed from Lake Ediza. (Photo by George Hafer.)

Figure 57. Glaciers at the south end of the Ritter Range, Sierra Nevada, July 1996.

> The inclination of the glacier is quite moderate at the head, and, as the sun had softened the névé, I made safe and rapid progress, running and sliding, and keeping up a sharp outlook for crevasses. About half a mile from the head, there is an ice-cascade, where the glacier pours over a sharp declivity and is shattered into massive blocks separated by deep, blue fissures. To thread my way through the slippery mazes of this crevassed portion seemed impossible, and I endeavored to avoid it by climbing off to the shoulder of the mountain. But the slopes rapidly steepened and at length fell away in sheer precipices, compelling a return to the ice. Fortunately the day had been warm enough to loosen the ice-crystals so as to admit of hollows being dug in the rotten portions of the blocks, thus enabling me to pick my way with far less difficulty than I had anticipated. Continuing down over the snout, and along the left lateral moraine, was only a confident saunter, showing that the ascent of the mountain by way of this glacier is easy, provided one is armed with an ax to cut steps here and there. (Muir, 1894, p. 57)

Do what Muir says, not what he did; carry an ice ax and a rope and do not visit glaciers alone.

LYELL GLACIER AND MACLURE GLACIER

Glaciers on Mounts Lyell and Maclure in Yosemite National Park are the most visited and studied of Sierra glaciers, in part because the popular John Muir Trail is close by (figs. 58, 59). These glaciers are located about 11 miles southeast of Tuolumne Meadows at the head of the Lyell Fork of the Tuolumne River. The glaciers are close together, and the Lyell Glacier, with surface area of about a quarter square mile, is the larger. The Lyell Glacier is divided by a rock ridge into two parts; most ascents of the mountain are over the western part. A bergschrund at the head of the glacier may be an obstacle to climbers in late summer.

John Muir visited both glaciers in 1871 and planted stakes to measure the rate of flow of the Maclure Glacier in August 1872. Joseph LeConte visited the Lyell Glacier with John Muir:

> [T]he main branch of the Tuolumne glacier, far up among the cliffs and peaks of Mt. Lyell, *still exists as a living glacier,* in a feeble state of vitality it is true, but certainly living. . . .
>
> The mass of snow occupying this cove is about a mile in length and about half a mile wide. Now, along the lower margin of this snow field and closely in contact with it, there is as perfect a *terminal moraine* as can be imagined. . . .
>
> Between the cliff . . . and the snow, there is an empty space like a crevasse, 4 to 5 feet wide, evidently produced by the tearing away of the moving snow from the perpendicular cliffs. In the language of the Alpine travelers, it is a *bergschrund.* (LeConte, 1873, p. 330)

Figure 58. Lyell Glacier, second largest in the Sierra Nevada, is accessible by the John Muir Trail from Tuolumne Meadows. (Photo by W. T. Lee, U.S. Geological Survey.)

Figure 59. Maclure Glacier, Sierra Nevada, vertical aerial photo, October 1966. North is to the right. The shadow of Mount Maclure covers much of the accumulation zone of the glacier. Look for the irregular firn line, crevasses, and curved flow bands. Glacier ice probably extends under the extensive lateral and end moraines. (Photo by W. V. Tangborn, U.S. Geological Survey.)

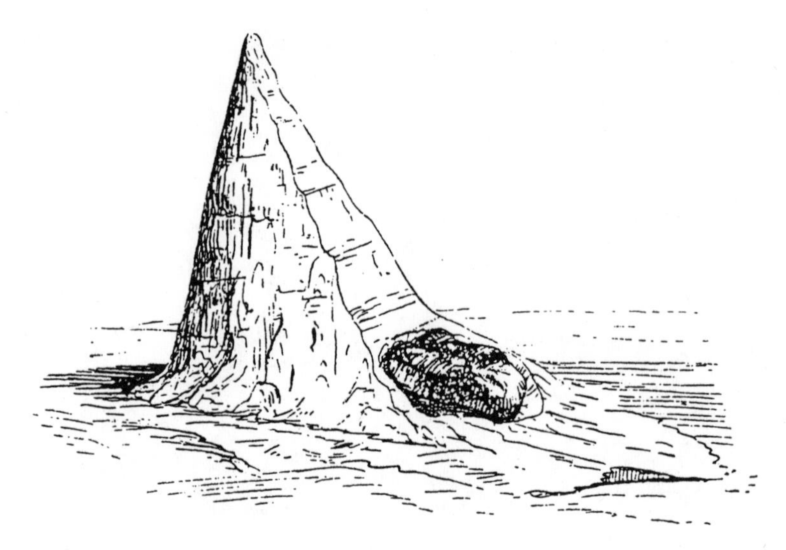

Figure 60. Ice pyramid on Mount Lyell. A small rock absorbs heat and melts underlying snow. Meltwater percolates down through the loose snow and refreezes in a pyramidal shape, which remains after surrounding loose snow has ablated. (Drawing by U.S. Geological Survey; Russell, 1885.)

Russell presents a map of the Lyell Glacier and describes interesting features such as ice pyramids and red snow (1885). Ice pyramids (fig. 60) are a few inches to several feet high and always have a stone a few inches across at their northern base. They form when a small stone

> lying on the surface of a glacier becomes heated and melts the porous ice beneath, and . . . the water thus formed freezes again into *compact* ice, which resists the sun's heat more effectually than the surrounding *porous* ice and hence is left as the general surface melts away. (Russell, 1885, p. 321)

A larger stone would form a glacier table rather than an ice pyramid because it would protect underlying snow rather than melt it. Red snow is formed by algae found in some snowbanks in late spring or summer, after melting has begun.

Willard D. Johnson descended into the bergschrund of the Lyell Glacier in an effort to understand how cirques are formed and how glaciers erode mountains. He wanted to observe basal sapping, the process he hypothesized was important in the development of cirques and glacial landscapes. The crevasse at the head of the glacier was 100 feet deep; at the bottom he found himself between ice on one side and the rock of the mountain wall on the other.

> The rock face, though hard and undecayed, was much riven, its fracture planes outlining sharply angular masses in all stages of displacement and dis-

> lodgment. Several blocks were tipped forward and rested against the opposite wall of ice; others, quite removed across the gap, were incorporated in the glacier mass at its base. Icicles of great size, and stalagmitic masses, were abundant; the fallen blocks in large part were ice-sheeted; and open seams in the cliff face held films of clear ice. (Johnson, 1904, p. 574)

The process whereby blocks of rock are quarried, or sapped, by a glacier is aided by meltwater entering the bergschrund and freezing in jointed rock. Expansion that accompanies freezing loosens and moves the joint blocks away from the cliff. Ice frozen both in the joints and the moving glacier provides the grip that permits the blocks of rock to be pulled away from the headwall of the cirque. Drawings of Johnson viewing displaced blocks of rock in the bergschrund of the Lyell Glacier, such as figure 61, have appeared in many geology textbooks.

In October of 1933 the mummified body of a mountain sheep, mostly intact, was found upright on a pedestal of ice on the Lyell Glacier, forming a glacier table of the rarest kind. Mountain sheep had been extinct in the Yosemite region for at least fifty years. The discoverers measured the rate of flow of the ice:

> We found that the glacier moved only one inch during a four-day period, or at the rate of seven and one-half feet per year. The ram was found 1936 feet from the head of the glacier. Now assuming that the animal fell or was caught in a slide while feeding on the crest of Mount Lyell and was buried in the bergschrund, it would take close to 250 years for the glacier to carry the sheep to the spot where found. (Beatty, 1933, p. 111)

Searching nearby they also found the remains of a marmot and a cony; for some reason both the sheep and the marmot lacked fur. It should be pointed out that the rate of flow of the glacier probably was not constant during the 250-year period mentioned in the quote. When the ice was thicker, as it was in the 1800s, the rate of flow would probably have been faster, which would reduce the length of time the sheep was in the ice.

Fran Hubbard described "ice worms" on the Lyell glacier:

> [a] large number of shallow, worm-shaped grooves in the thin crust of two-day-old snow. At the end of each wiggly groove was a minute object, but no two were the same. Of the seven examined, three were small bits of rock, two small pieces of dirt, and two, small dead insects. Colors ranged from black (the dirt) to almost white (the insects). The grooves followed no set pattern but ran in all directions, uphill and down. They varied in length from two to more than six inches. The only constant factor was the presence of the small foreign body at one end.
>
> Was the wind responsible for the formation of these "ice worms"? If so, why did no two of them run in the same direction? Angle of slope seemed to

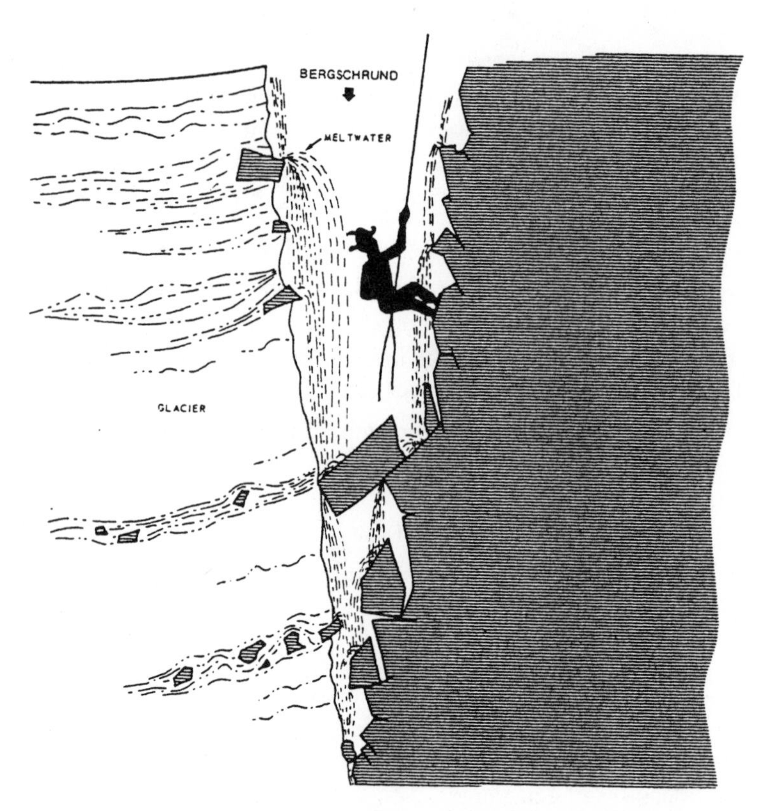

Figure 61. A drawing inspired by W. D. Johnson's descent into the bergschrund of the Lyell Glacier. When the crevasse is filled, ice in contact with the cliff plucks jointed rock from the mountainside. By this means a glacier erodes headward, creating and enlarging its cirque. (Drawing by Ron Morgan, used with permission of the California Department of Conservation, Division of Mines and Geology; Hill, 1974.)

> have no apparent bearing on the pattern. Was the sun the cause? If so, why didn't the particles melt down out of sight, as larger objects have done on the glacier's surface? Perhaps it was a combination of the two. Or perhaps this will prove to be another unexplained mystery of movement.
>
> (Hubbard, 1954, p. 64)

The "ice worms" remain unexplained. Dare we hypothesize that ice worms ate the fur off the mummified sheep and marmot? Strangely, small worms do live in the ice of some glaciers in North America (Ferguson, 1992; Sharp, 1988), but none has been found in California glaciers to my knowledge.

The east part of the Lyell Glacier is known to have retreated 87 feet between 1931 and 1939 (Matthes, 1942). Harrison presents photos of the Lyell and Maclure Glaciers along with a detailed discussion of changes in size that have been observed since their first study (1950, 1951). He also presents photos of Lyell Glacier from the same location in different years, which can be used to assess changes in the ice cover (1960); significant retreat of the terminus occurred between 1937 and 1953.

DANA GLACIER

The Dana Glacier is on the north slope of Mount Dana, a 13,000-foot peak near Tioga Pass on the eastern boundary of Yosemite National Park (fig. 62). A prominent feature of the Dana Glacier is the large couloir that rises to a ridge near the summit of the mountain. Russell says the glacier is 2,000 feet long in the direction of flow, and presents three fine drawings of it (1885). Harrison describes changes observed on Dana Glacier between 1883 and 1949, including two rock slides that covered much of the ice, and he gives good photographs of the glacier (1950, 1951, 1960). Huber reports the Dana Glacier lost three-fourths of its surface area and much of its volume in the sixty-seven years between 1908 and 1975 (1987).

Figure 62. Dana Glacier, Sierra Nevada, with prominent couloir, bergschrund, crevasses, and moraine. (Photo by I. C. Russell, U.S. Geological Survey.)

Figure 63. Conness Glacier, Sierra Nevada, viewed from Steelhead Lake, July 1996. The prominent crevasse is the bergschrund, which separates the glacier from nonflowing ice in the several couloirs.

CONNESS GLACIER

The Conness Glacier (fig. 63) is located on the northeast side of Mount Conness, a 12,590-foot peak on the crest of the Sierra Nevada on the east boundary of Yosemite National Park. Matthes tells how, in 1939, a large lake of meltwater formed behind the terminal moraine and, overflowing, eroded a vertical trench that revealed that the terminal moraine consisted of a core of ice covered with till (1948). Matthes points out that other glaciers in the Sierra and elsewhere also have lakes behind their terminal moraines and suggests that ice-cored moraines might be quite common, for without ice the till of the moraine would likely be permeable and unable to contain the water. This observation is important with respect to the existence of "debris-covered glaciers" discussed later in this chapter.

MATTHES GLACIERS

Several glaciers on Glacier Divide along the north boundary of Kings Canyon National Park, in the Sierra Nevada, have been named the Matthes Glaciers in honor of François Matthes of the U.S. Geological Survey. Six of the glaciers have dimensions greater than a quarter mile; several others are smaller. All are at an elevation of about 11,800 feet.

GLACIERETS NEAR MOUNT ABBOT

Mount Abbot (13,704 feet) is in the vicinity of Rock Creek and the Mono Recesses, between Yosemite and Kings Canyon National Parks. Joseph Birman describes glacierets there:

> The latest pulse of the Little Ice Age . . . is represented . . . by at least 13 cliff ice masses, perennial snow accumulations, or glacierets, a few of which contain bergschrund. Of those mapped, the smallest is about 500 feet across; the largest, in Rock Creek on the pedestal of Mt. Mills, Mt. Abbot, and Mt. Dade, is about 4,500 feet as measured parallel to the slope, and from 500 to 1500 feet in down-valley extent. Altitude of the lowest mapped is 10,500 feet in the First Recess; the highest is 12,500 feet at the head of Rock Creek.
>
> Slopes are steep, and the glacierets are against the cirque headwalls. In late August 1951 the Recess Peak glacieret had 5 feet of loose granular snow on the surface, underlain by 5–8 feet of hard-packed granular snow, underlain in turn by clear ice. In 1951 the toes of all glacierets observed by the writer were not in equilibrium with the surrounding depositional material but were several hundred feet back and were separated by deep fosses.
>
> (Birman, 1964, p. 53)

This is a good description of the change from old snow to firn to glacier ice.

GLACIERETS OF THE TRINITY ALPS

Robert Sharp describes two glacierets in the Trinity Alps of the Klamath Mountains of northwest California:

> Two small vestigial bodies of glacier ice remain in the area. One lies at the head of Grizzly Creek on the north side of Thompson Peak (8936 feet). The other occupies an adjacent cirque at the head of Thompson Creek. . . .
>
> Such bodies cover 5 or 6 acres, lie at elevations between 8200 and 8500 feet, display crevasses, and consist of somewhat dirty ice in crystals up to one inch in diameter. When visited on September 1, 1956, a year of considerable residual snow, they were mostly snow-covered. Views from a distance in other, drier years, suggest that at times 30 to 40 percent of the ice may be exposed during summer. A short distance beyond the margins of both bodies are fresh bouldery moraines indicative of greater size and activity in the immediate past. (Sharp, 1960, p. 337)

These glacierets are over 2,000 feet lower than those of the Sierra Nevada, a condition made possible by a more northerly latitude and proximity to the coast.

ROCK GLACIERS AND DEBRIS-COVERED GLACIERS

There are dozens, perhaps hundreds, of interesting landforms called rock glaciers, or rock streams, in the higher elevations of the Sierra Nevada (fig. 64), and at least one is known on Mount Shasta. Rock glaciers are

> tongue-shaped accumulations of rubble and blocks. In composition and texture they resemble the numerous talus cones near by. The rubble and blocks of the rock streams can thus be expected to have originated in the same manner as the similar material deposited in the talus cones, namely through common weathering of the neighboring cliffs. The form assumed after deposition, a tongue shape in the case of rock stream, a cone shape in that of talus, is the only difference between the two deposits. (Kesseli, 1941, p. 204)

Rock glaciers range from a few hundred feet to a mile long, and from 10 to 200 feet thick. They often are present in cirques below cliffs of highly jointed

Figure 64. Rock glacier on Cardinal Mountain, Sierra Nevada, southwest of Big Pine, Aug. 24, 1992. Curved ridges caused by flow and steep slopes at the advancing foot are characteristic of active rock glaciers. (Photo by Douglas H. Clark.)

rock from which rockfall is rapid. Kesseli reviewed various hypotheses for the origin of rock glaciers and concluded they are related to true glaciers because they occur where glaciers formed in the past, unlike simple masses of talus that form at diverse locations. In the Mount Pinchot quadrangle of the Sierra Nevada,

> [r]ock glaciers occur in the high cirques on north-to-northeast-facing slopes. The lower end . . . is a steep front of large, angular blocks resting at about the angle of repose. The average elevation of the base of this front is about 11,500 feet. The rock glaciers are from one-half to one-fourth mile in length and are fed by talus cones at their upper end. On the upper surface, parallel to the lower end and convex down-canyon are parallel ridges 10 to 15 feet high and about five times as far from crest to crest. The upper end of the rock glacier is ill defined and merges with talus, ice, or bedrock at the base of the cirque headwall. (Moore, 1963, p. 140)

The shape and the convex-downhill concentric ridges, or wrinkles, that most have, indicate that the mass of rubble has flowed downhill, hence the word glacier in the name.

Douglas H. Clark, Malcolm Clark, and Alan Gillespie report that some rock glaciers in the Sierra Nevada are true glaciers with a veneer of talus only a few feet thick (fig. 65). They studied the Southfork Pass rock glacier, located near the Middle Palisade Glacier on Big Pine Creek. It is about a mile long and 900 feet wide. They compared aerial photos taken at various times, observed "clear glacier ice" beneath a few feet of debris exposed on the wall of a meltwater pond, and dug a pit to see what was under the debris.

> In combination, our observations demonstrate that Southfork Pass rock glacier is glacial in origin, is cored by glacier ice, and is therefore described more accurately as a debris-covered glacier. . . . We emphasize that, although unusually large, this debris-covered glacier is morphologically, texturally, and structurally indistinguishable from most other active valley-floor rock glaciers in the Sierra Nevada. (Clark, Clark, and Gillespie, 1994, p. 143)

The Southfork Pass debris-covered glacier is advancing while the adjacent Middle Palisade Glacier, despite being higher and having a larger accumulation area, is retreating.

> The insulation provided by continuous debris cover on a glacier clearly allows an ablation area to extend to lower elevations than those of equivalent bare-ice glaciers. Debris-covered glaciers and ice-cored moraines may thus persist well after climatic change thins or destroys bare glacier ice.
> (Clark, Clark, and Gillespie, 1994, p. 147)

Figure 65. Meltwater pond, exposure of ice, and surficial boulders and cobbles on a debris-covered glacier (or ice-cored rock glacier). The thin cover of rock debris protects underlying ice from melting and permits debris-covered glaciers to exist at lower elevations than bare-ice glaciers. Southfork Pass rock glacier, Big Pine Creek, Sierra Nevada, Aug. 25, 1992. (Photo by Douglas H. Clark.)

Because debris-covered glaciers behave differently from regular glaciers, studies of regional climate will be invalid if the elevations of deposits from debris-covered glaciers are inadvertently utilized along with those of regular glaciers.

Sarah Konrad and Doug H. Clark find evidence that some rock glaciers formed more than 1,000 years ago and so are older than existing (regular) glaciers of the Sierra Nevada (1996). These rock glaciers may be the only evidence of glaciers in the Sierra in the approximate 9,300 years between the end of the Pleistocene and the birth of the Matthes (Little Ice Age) glaciers.

THE FUTURE OF CALIFORNIA GLACIERS

What will happen to California glaciers in the future? No one can predict, so we must learn by observing. Unfortunately it takes a year to get a year's worth of observation, so we must be patient. We have many good aerial photographs, maps, and photographs taken from known locations of California glaciers for reference and comparison in decades and centuries to come. Equally important is the existence of institutions capable of long-term monitoring, such as the U.S. Geological Survey, the California Division of Mines and Geology, the National Park Service, and various colleges and universities.

Natural climate changes have made California glaciers advance and retreat many times in the last 1.5 million years. Now there is another variable: global warming, climate change that possibly is occurring because of human alteration of the atmosphere caused by burning fossil fuels and deforestation. It remains to be seen whether such warming is real, how it might interact with natural changes, and what the final result will be for California.

While waiting it is fun to speculate. Perhaps global warming will counter the effect of the next natural Ice Age, thereby preventing the destruction of civilization (air pollution is good!). Perhaps a warming planet will cause all California glaciers to disappear (air pollution is bad!). Perhaps warming of the Pacific Ocean will result in more evaporation of water, creating, for California, cloudy and cooler summers, conditions that will permit our glaciers to grow to what they were 20,000 years ago. Our great-great-great-grandchildren might see another 60-mile-long Tuolumne Glacier, see ice again a thousand feet thick in Yosemite Valley, and live in California, newly proclaimed by a future governor to be "The Glacier State."

REFERENCES CITED

Beatty, M. E. 1933. Mountain sheep found in Lyell glacier. *Yosemite Nature Notes* 12: 110–12.

Birman, Joseph H. 1964. *Glacial geology across the crest of the Sierra Nevada, California.* Geol. Soc. of Amer. Special Paper 75.

Burnett, John L. 1964. Glacier trails of California. *Mineral Info. Service* 17: 33.

Carter, Ernest S. 1984. Whitney Glacier, 1983 . . . record of a climb. *Calif. Geology* 37: 3–8.

Christiansen, Robert L. 1976. Volcanic evolution of Mt. Shasta, California. *Geol. Soc. of Amer. Abstracts with Programs* 8: 360–61.

Clark, Douglas H., Malcolm M. Clark, and Alan R. Gillespie. 1994. Debris-covered glaciers in the Sierra Nevada, California, and their implications for snowline reconstructions. *Quaternary Research* 41: 139–53.

Clark, Douglas H., Alan R. Gillespie, and Malcolm M. Clark. 1993. Effects of local topography on ELA trends of cirque glaciers. *Geol. Soc. of Amer. Abstracts with Programs* 25: A156.

Diller, J. S. 1895. Mount Shasta, a typical volcano. *National Geographic Society Monographs* 1: 237–68.

Driedger, Carolyn L., and Paul M. Kennard. 1986. *Ice volumes on Cascade volcanoes: Mount Rainier, Mount Hood, and Mount Shasta.* U.S. Geol. Survey Prof. Paper 1365.

Ferguson, Sue A. 1992. *Glaciers of North America: A field guide.* Golden, Colo.: Fulcrum Publishing.

Harrison, A. E. 1950. Glaciers then and now. *Sierra Club Bull.* 35: 113–16.

———. 1951. Are our glaciers advancing? *Sierra Club Bull.* 36: 78–81.

———. 1960. *Exploring glaciers with a camera.* San Francisco: Sierra Club Books.

Hershey, Oscar H. 1903. Some evidence of two glacial stages in the Klamath Mountains in California. *American Geologist* 31: 139–56.

Hill, Mary. 1974. Glaciers—a picture story. *Calif. Geology* 27: 23–45.

———. 1975. Living glaciers of California. *Calif. Geology* 28: 171–77.

———. 1978. Whitney Glacier: Record of a climb. *Calif. Geology* 31: 10–14.

Hill, Mary, and Elisabeth L. Egenhoff. 1976. A California jökulhlaup. *Calif. Geology* 29: 154–58.

Hubbard, Fran. 1954. Lyell Glacier's mysterious "ice worms." *Natural History* 63: 64–65.

Huber, N. King. 1987. *The geologic story of Yosemite National Park.* U.S. Geol. Survey Bull. 1595.

Johnson, W. D. 1904. The profile of maturity in alpine glacial erosion. *Journal of Geology* 12: 569–78.

Kesseli, John E. 1941. Rock streams in the Sierra Nevada, California. *Geographical Review* 31: 203–27.

King, Clarence. 1872. *Mountaineering in the Sierra Nevada.* Boston: James R. Osgood and Co.

Konrad, Sarah K., and Doug H. Clark. 1996. Multiple neoglacial advances recognized in Sierra Nevada rock glaciers. *Geol. Soc. of Amer. Abstracts with Programs* 28, no. 7: 433.

LeConte, Joseph. 1873. On some of the ancient glaciers of the Sierras. *American Jour. of Science,* 3rd ser., 5: 325–42.

Leonard, Richard M. 1951. A climber's guide to the High Sierra. *Sierra Club Bulletin* 36: 126–32.

Matthes, F. E. 1942. Glaciers. In *Physics of the Earth,* edited by O. E. Meinzer. Part 9, Hydrology, 149–219. New York: McGraw-Hill.

———. 1948. Moraines with ice-cores in the Sierra Nevada. *Sierra Club Bull.* 33: 87–96.

———. 1962. *François Matthes and the marks of time.* Edited by F. Fryxell. San Francisco: Sierra Club.

Miller, C. Dan. 1980. *Potential hazards from future eruptions in the vicinity of Mount Shasta volcano, northern California.* U.S. Geol. Survey Bull. 1503.

Moore, James G. 1963. *Geology of the Mount Pinchot quadrangle, southern Sierra Nevada, California.* U.S. Geol. Survey Bull. 1130.

Muir, John. [1894] 1961. *The mountains of California.* Garden City, N.Y.: Doubleday and Co.

Pierce, Bob, and Margaret Pierce. 1973. *Merced Peak: High Sierra Hiking Guide No. 11.* Berkeley: Wilderness Press.

Raub, W. D., A. Post, C. S. Brown, and M. F. Meier. 1980. Perennial ice masses of the Sierra Nevada, California. *Proceedings of the International Assoc. of Hydrological Sci.,* no. 126: 33–34.

Rhodes, Philip T. 1987. Historic glacier fluctuations at Mount Shasta, Siskiyou County. *Calif. Geology* 40: 205–11.

Russell, Israel C. 1885. *Existing glaciers of the United States.* U.S. Geol. Survey 5th Annual Report, 1883–1884, 303–55.

Schaffer, Jeffrey P. 1992. *Yosemite National Park.* 3rd ed. Berkeley: Wilderness Press. 1994 update.
Selters, Andy, and Michael Zanger. 1989. *The Mt. Shasta book.* Berkeley: Wilderness Press.
Sharp, Robert P. 1960. Pleistocene glaciation in the Trinity Alps of northern California. *American Jour. of Science* 258: 305–40.
———. 1988. *Living ice: Understanding glaciers and glaciation.* Cambridge: Cambridge Univ. Press.
Trent, D. D. 1983. The Palisade Glacier, Inyo County. *Calif. Geology* 36: 264–69.
Von Engeln, O. D. 1933. Palisade Glacier of the High Sierra of California. *Geol. Soc. of Amer. Bull.* 44: 575–600.

CHAPTER EIGHT

TREES, LAKES, POLLEN, AND PACKRATS

Glacial deposits are not the only evidence of climate changes that cause advance and retreat of glaciers. Other studies are being done in California that shed light on climate and indirectly on glaciation. Some of these studies will be described to illustrate how they relate to glacial history.

TREES

That the snow line on Mount Shasta was higher, and the climate different, at some time prior to the late 1800s was noted by Clarence King on the moraine of the McCloud Glacier:

> Here and there, half buried in the drift, we came across the tall, noble trunks of avalanche-killed trees. In comparing their straight, symmetrical growth with the singularly matted condition of the living dwarfed trees, I find the indication of a great climatic change. (King, 1872, p. 262)

There are many very long-lived trees found in the White-Inyo Mountains. The elevation of trees, living or dead, and their age as revealed by tree-rings can be used to establish the elevation of former timberlines:

> Remains of dead bristlecone pine . . . are found at altitudes up to 150 m above present treeline in the White Mountains. Standing snags and remnants in two study areas were mapped and sampled for dating by tree-ring and radiocarbon methods. The oldest remnants represent trees established more than 7400 y.a. [years ago]. Experimental and empirical evidence indicates that the position of the treeline is closely related to warm-season temperatures, but that precipitation may also be important in at least one of the areas. The upper treeline was at high levels in both areas until after about 2200 B.C., in-

> dicating warm-season temperatures about 3.5°F higher than those of the past few hundred years. However, the record is incomplete, relative warmth may have been maintained until at least 1500 B.C. Cooler and wetter conditions are indicated for the period 1500 B.C.–500 B.C., followed by a period of cool but drier climate. A major treeline decline occurred between about A.D. 1100 and A.D. 1500, probably reflecting onset of cold and dry conditions. High reproduction rates and establishment of scattered seedlings at high altitudes within the past 100 yr. represents [sic] an incipient treeline advance, which reflected a general climatic warming beginning in the mid-19th century that has lasted until recent decades in the western United States. This evidence for climatic variation is broadly consistent with the record of Neoglacial advances in the North American Cordillera, and supports Antevs' concept of a warm "altithermal age" in the Great Basin. (LaMarche, 1973, p. 632)

Trees can provide high resolution information. Louis Scuderi studied a 1,200-year record of foxtail pine in the southern Sierra Nevada:

> Minima in the ringwidth record, reflecting marked temperature declines occurred at 810, 1470, 1610, 1700, and 1810. Cold periods of lesser extent are also indicated between 1190 to 1400 and suggest that the initial pulses of the Sierran Matthes advances may have begun as early as 1190, or 150 yr. earlier than previously dated. (Scuderi, 1987, p. 220)

These intervals of alternately colder and warmer years during an overall colder period agrees well with the depiction of conditions during the Little Ice Age by Grove (1988), quoted in chapter 6 of this book. Scuderi also studied trees in the southern Sierra Nevada to see if volcanic eruptions are detectable in the tree record:

> Ringwidth variations from temperature-sensitive upper timberline sites in the Sierra Nevada show a marked correspondence to the decadal pattern of volcanic sulfate aerosols recorded in a Greenland ice-core acidity profile and a significant negative growth response to individual explosive volcanic events. The appearance of single events in the mid-latitude tree-ring record, in connection with ice-core evidence from the arctic and historical records from the Mediterranean, indicates that the majority of these events represent climatically effective volcanic eruptions, producing temperature decreases on the order of 1°C for up to 2 yr. after the initial eruption. Clusters of climatically effective volcanic events may serve as a trigger to glaciation and are consistently associated with lowered ringwidths and late-Holocene glacier advance in the Sierra Nevada. (Scuderi, 1990, p. 67)

Lisa Graumlich also studied foxtail pine in the southern Sierra Nevada:

> Tree-ring data from subalpine conifers . . . were used to reconstruct temperature and precipitation back to A.D. 800. . . . The summer temperature reconstruction shows fluctuations on centennial and longer time scales including a period with temperatures exceeding late 20th-century values from ca. 1100 to 1375 A.D., corresponding to the Medieval Warm Period identified in other proxy data sources, and a period of cold temperatures from ca. 1450 to 1850 corresponding to the Little Ice Age. (Graumlich, 1993, p. 249)

Her paper concludes,

> [R]esults presented here indicate that the high precipitation levels of the mid-20th century have occurred only three times in the previous 1000-yr. record. These results suggest that current drought in California is not an anomaly when considered in a long-term context and that agricultural, industrial, and urban systems that are dependent on water resources are highly vulnerable to disruption. (Graumlich, 1993, p. 255)

This is ominous information for the most populous state in the nation, the population of which is predicted to increase by over 10 million in the next twenty years.

LAKES

Mono and Owens Lakes east of the Sierra Nevada are valuable for climate studies because, being in the Great Basin, which lacks drainage to the ocean, lake levels depend on runoff from adjacent mountains.

> Direct precipitation in the form of rain is normally insignificant on the east flank of the range and in Owens Valley, which has justly earned the name of "the land of little rain." The same abundant snows that feed the lake also maintain the fifty-odd small glaciers on the range. It can hardly be doubted that the large ancient Owens Lake was contemporaneous with the great Pleistocene glaciers of the Sierra Nevada, some of which attained lengths of 30 to 60 miles; neither can it be reasonably doubted that when the temperature rose and the snow on the range was so reduced that it failed to maintain the ancient lake in existence, it likewise failed to maintain the glaciers.
> (Matthes, 1942, p. 212)

Thus it is reasoned that establishing former elevations of these lakes, and the time that each level was reached, will relate to what glaciers in the adjacent mountains were doing.

Lake sediments are potentially a much better record of glacial advances and retreats than moraines are. There are a number of reasons why this is so. Whereas moraines accumulate horizontally, deposits in water accumulate vertically, and this makes a lot of difference. A large glacier advance (horizontal) might completely destroy all evidence of previous glaciations by pushing aside older moraines and incorporating their till into its own. In contrast, sediment deposited in lakes accumulate as a layered sequence (vertical) and only rarely does deposition of a layer, big or small, alter what went before. In addition, layered sediments show time sequence clearly, but recreating the order in which a cluster of moraines formed can be tricky. Complete rock-records are rarer on land than under water. For example, volcanic eruptions commonly cause distinctive layers of ash to fall over very large areas, and these layers are of great value in dating geologic events. The ash is much more likely to be preserved in a lake than on land.

SELECTED STUDIES AT MONO LAKE

Mono Lake has a record as much as 170,000 years old (Lajoie, 1968). Kenneth Lajoie and Stephen Robinson studied 10 meters of glacio-lacustrine silt and identified high lake levels at 36–34 thousand years ago, 28–24 thousand years ago, and 14–12 thousand years ago. (In the quote that follows, "ka" is short for "thousand years" and "BP" means "before present.")

> The last brief highstand (14–12 ka BP) correlates well with the last lacustrine highstands in other closed basins throughout the Great Basin, but does not correlate with the last Sierran glacial maximum (Tioga), which presumably was contemporaneous with the last continental glacial maximum and its related sea-level lowstand (both 19–17 ka BP). . . . Mono Lake, which was fed directly by glacial runoff, stood at an intermediate level . . . during the glacial maximum. To the north, glacier-fed Lake Lahontan also stood at an intermediate level during this period of time. Thus, the climatic factors that produced maximum glacial conditions did not produce maximum lacustrine conditions. . . . The 4–7 ka lag of the lacustrine maximum behind the glacial maximum may indicate that the increase in temperature at the end of the Pleistocene was not accompanied by a decrease in precipitation.
>
> (Lajoie and Robinson, 1982, p. 179)

This is a good example of how studying both lakes and glaciers produces insights that study of either alone might not provide. In this case, it appears that the end of the Ice Age was accompanied by warm and wet conditions, a circumstance that would explain how the lake might continue to become larger for 4,000 to 7,000 years after the glacier maximum.

Newton, Lund, and Davis found that Mono Lake was low for at least 1,000 years prior to 5,500 years ago (1987). This is the warm time of the Altithermal. They also noted, "Sometime around 500 years BP the lake suddenly rose again to a level at least 16 m higher than the present." This is the time of the Little Ice Age. Finally they found the lake was high between 5,500–2,900 years ago. Does this highstand correspond with an unrecognized glacial advance in the Sierra Nevada?

Stine found very high lake levels at 3,770 years ago and at 300 years ago (1990). The older highstand agrees with the study just discussed, and it also suggests a possible unrecognized Sierra glaciation. The second exceptional high, 300 years ago, is one of five highstands during the last 900 years that reflect climatic fluctuation during the Little Ice Age.

SELECTED STUDIES AT OWENS LAKE

Sediments of Owens Lake and moraines of the Sierra Nevada have been compared with the timing of Heinrich events (expulsions of icebergs into the North Atlantic Ocean) (Benson and others, 1996; Phillips and others, 1996). The purpose of the studies was to see if the icebergs and glacial maxima in the Sierra occurred at the same times. If they did, it would confirm a hypothesized connection between cooling cycles in the North Atlantic and glacier advances in western North America.

> [A]lthough expansions of Sierra Nevada glaciers were apparently related to Heinrich events, the relation was probably not one of simple synchroneity. . . . Although the nature of the climatic linkages remains to be explained, the close association of both glacial and lacustrine fluctuations in western North America with these events supports the hypothesis that they were of global scale. (Phillips and others, 1996, p. 750)

POLLEN

Plants are sensitive to climate, and the vegetation present in an area in past times can be determined from the kinds of pollen found in meadows, bogs, swamps, and lakes. Pollen records are not capable of such high resolution as tree rings but can provide information about average temperature and precipitation. Three studies of pollen from different locations in the Sierra Nevada found a significant moist period beginning between 3,000 and 4,500 years ago (Davis and others, 1985; Anderson and Smith, 1994; and Edlund and Byrne, 1990). These pollen studies agree with the 4,000-year-old age of Owens Lake that Matthes used to estimate the beginning of the Little Ice Age (as he used the term) and highstands of Mono Lake at 3,770 years ago (Stine,

1990) and 2,900–5,500 years ago (Newton, Lund, and Davis, 1987). There is no direct evidence of glaciation known in the Sierra Nevada from about 3,500–4,500 years ago, but the absence may be caused by destruction of moraines by the Matthes glaciers. Lake and pollen may both record a Holocene glaciation of the Sierra Nevada, the moraines of which have been destroyed.

PACKRATS

Packrats not only steal from backpackers their meager supply of food and valuables, they also accumulate vegetation from the vicinity of where they live, and it may become preserved in their middens. *Midden* is a polite name for manure pile or refuse heap. A packrat midden can remain intact for thousands of years and can be studied to determine the vegetation that was present in the area at the time it was formed. Steven Jennings and Deborah Elliott-Fisk explain how middens are studied:

> General characteristics of eight middens from six sites in the White Mountain region were recorded. These data include substrate, azimuth, slope angle, and topographic position. All plant species within 30 m of the midden site (the nominal foraging range of *Neotoma* spp.) were recorded. . . .
>
> The material in each midden was prepared by removing the outer weathering rind with a chisel. A portion of the midden was reserved for future analysis. The remaining midden was dissolved in water to remove the uriniferous matrix. The aqueous mixture was screened and washed using standard number-20 soil sieves. Organic material that remained was then oven-dried at 80°C for 24 hr. Plant macrofossils were identified using herbarium voucher specimens. The frequency of plant taxa in the midden was also reported on a semiquantitative basis. . . . Fecal pellets were radiocarbon dated to preserve plant macrofossils for other analyses. (Jennings and Elliott-Fisk, 1993, p. 215)

One of their middens had a radiocarbon age of 19,290 years old. None of the plant species in that midden is present at the site today, and the vegetation that was preserved in the midden "indicates 60 m elevational depression" 19,000 years ago, a time of cooler and moister climate near the end of the Tioga glaciation.

REFERENCES CITED

Anderson, R. Scott, and Susan J. Smith. 1994. Paleoclimatic interpretations of meadow sediment and pollen stratigraphies from California. *Geology* 22: 723–26.

Benson, Larry V., James W. Burdett, Michaele Kashgarian, Steve P. Lund, Fred M. Phillips, and Robert O. Rye. 1996. Climatic and hydrologic oscillations in the Owens Lake basin and adjacent Sierra Nevada, California. *Science* 274: 746–49.

Davis, Owen K., R. Scott Anderson, Patricia L. Fall, Mary K. O'Rourke, and Robert S. Thompson. 1985. Palynological evidence for early Holocene aridity in the southern Sierra Nevada, California. *Quaternary Research* 24: 322–32.

Edlund, Eric G., and A. Roger Byrne. 1990. Reconstruction of late-Quaternary vegetation and climatic change in the central Sierra Nevada, California. *Geol. Soc. of Amer. Abstracts with Programs* 22, no. 3: 20.

Graumlich, Lisa J. 1993. A 1000-year record of temperature and precipitation in the Sierra Nevada. *Quaternary Research* 39: 249–55.

Grove, Jean M. 1988. *The little ice age.* London: Methuen.

Jennings, Steven A., and Deborah Elliott-Fisk. 1993. Packrat midden evidence of late Quaternary vegetation change in the White Mountains, California-Nevada. *Quaternary Research* 39: 214–21.

King, Clarence. 1872. *Mountaineering in the Sierra Nevada.* Boston: James R. Osgood and Co.

Lajoie, K. R. 1968. Late Quaternary stratigraphy and geologic history of Mono Basin, eastern California. Ph.D. diss., Univ. of Calif. at Berkeley.

Lajoie, K. R., and S. W. Robinson. 1982. Late Quaternary glacio-lacustrine chronology, Mono Basin, California. *Geol. Soc. of Amer. Abstracts with Programs* 14: 179.

LaMarche, Valmore C., Jr. 1973. Holocene climatic variations inferred from treeline fluctuations in the White Mountains, California. *Quaternary Research* 3: 632–60.

Matthes, F. E. 1942. Glaciers. In *Physics of the earth,* edited by O. E. Meinzer. Part 9, Hydrology, 149–219. New York: McGraw-Hill.

Newton, Mark S., Steve P. Lund, and Owen K. Davis. 1987. Magnetostratigraphy of sediment cores from Mono Lake, CA, and dating of Holocene climatic events. *Geol. Soc. of Amer. Abstracts with Programs* 19: 787.

Phillips, Fred M., Marek G. Zreda, Larry V. Benson, Mitchell A. Plummer, David Elmore, and Pankaj Sharma. 1996. Chronology for fluctuations in late Pleistocene Sierra Nevada glaciers and lakes. *Science* 274: 749–51.

Scuderi, Louis A. 1987. Glacier variations in the Sierra Nevada, California, as related to a 1200-year tree-ring chronology. *Quaternary Research* 27: 220–31.

———. 1990. Tree-ring evidence for climatically effective volcanic eruptions. *Quaternary Research* 34: 67–85.

Stine, S. 1990. Past climate at Mono Lake. *Nature* 345: 391.

PART 4

SEEING FOR YOURSELF

CHAPTER NINE

A FIELD TRIP

THE SIERRA GLACIER TOUR

You can visit places where you can see existing glaciers and glacial landforms for yourself. The outing described here, "The Sierra Glacier Tour," is a road trip about a hundred miles long, mostly in Yosemite National Park. The tour begins, appropriately, at Glacier Point in Yosemite National Park; places en route include Yosemite Valley, Tuolumne Meadows, Tioga Pass, Saddlebag Lake, and Lee Vining Canyon. The tour ends at the junction of Highways 120 and 395 near Lee Vining and Mono Lake. All of the places on the tour are scenic and worthy destinations with things to do and much to see in addition to the glacial subjects that are the focus of the trip. Bring binoculars for sure and camera and fishing tackle if you wish.

The time to take the trip is between mid-June and late September because the Glacier Point Road and the Tioga Pass Road are closed by snow much of the year. Opening and closing dates depend on snowfall, not the calendar; persons planning a trip in early June or in October should first establish that roads are in fact open.

Most stops on the tour involve only short walks, others are day hikes of several miles round-trip, and two are overnight backpack trips. Persons new to hiking or backpacking in high mountains should consult a how-to-do-it book and/or invite someone with experience to share the trip. Some detailed travel instructions are given in the following pages, but the focus is on where to go and what to look for rather than how to get there or where to eat and sleep. You should have a California road map and a Yosemite Park map. Yosemite National Park is usually crowded, and reservations will probably be necessary for campgrounds or other overnight facilities. Call or write

Yosemite National Park for information about making reservations. Plan well ahead if possible, but it is worth trying for reservations even at the last moment, as cancellations do occur. Weekends are busier than weekdays. Some campgrounds within the park are first come, first served, no reservations accepted; assume that they, as well as Forest Service campgrounds outside the park, will be filled by noon. There are hotels, motels, and campgrounds along highways east and west of Yosemite Park.

VISITING GLACIERS

Most people will find it satisfying enough to walk where Ice Age glaciers once flowed and inspect existing glaciers from a distance with the aid of binoculars. But for those who crave to walk on ice, WARNING: glaciers can be dangerous, and visiting them is not for the casual hiker. Aspiring glacier explorers should read a guide to mountaineering equipment and techniques, such as chapters 9 and 10 of Sue Ferguson's *Glaciers of North America: A Field Guide.* Shasta Mountain Guides, with offices in the city of Mount Shasta, offer on-the-ice courses in glacier travel, safety, and rescue. They also offer guided visits to the glaciers, as well as summit climbs.

Never visit a glacier alone, and if possible go with someone with experience. Carry rain gear and special equipment such as climbing rope, ice ax, protective eyewear, and possibly crampons. Simply getting onto a glacier is a potential hazard because there is often a marginal moraine with unstable rocks and boulders on steep slopes. Snow cover makes walking or skiing on a glacier easy in winter or early spring when snow may bridge crevasses, but in summer the snow bridges vanish or become too thin to support a person. While it is true that people have walked across glaciers and crevasses without even knowing it, others have fallen through snow bridges and been injured or killed. If you encounter an isolated patch of snow on ice, stay off the snow; it may exist there after the surrounding snow is gone because it is located over the cold air in a crevasse. Icefalls, such as those on the Whitney Glacier of Mount Shasta, are only for experienced and well-equipped mountaineering parties.

D. D. Trent, of Citrus College, California, has studied the Palisade Glacier in the Sierra Nevada for many years and led numerous groups onto the ice. Here is his advice:

> An important point that I like to make when discussing doing research on California glaciers is that it's a bit arduous. They are all at high elevations in Wilderness Areas, which translates into no helicopter support, and long backpacks carrying heavy loads in thin air. Enormous efforts go into collecting data. . . . [W]e experienced intense sun, fog, rain, sleet and snow (in early

Sept.). A rockfall just missed my 14-year-old son in a roped-up party but hit one of my students and broke his hip.

We ran into an experienced high elevation climber (who had bagged peaks in the Andes and elsewhere) on one trip who was very tired, had raspy breathing, etc., and was clearly in the early stages of pulmonary edema—we got him off the mountain fast. We also dealt with two cases of hypothermia, not in our party, but some young people who had read a lot about the wonders of high elevation Sierra but were clearly unprepared for the afternoon cool rain and wind chill effects. Another party, two young women who were enraptured by the glories of the High Sierra, camped near us. Together we endured a night of terrific rain, awesome thunder and lightning bolts striking around us while we were out of our tents in our BVDs keeping a low profile in a boulder field. The women were up early the next morning, leaving as fast as they could for the low country; remaining behind were their stoves, wet sleeping bags, tent, etc. (D. D. Trent, letter, 22 August 1996)

"Gentle Wilderness" is an appropriate name for the High Sierra most of the time, but with little warning it can turn lethal for the unprepared, inexperienced, or reckless.

GLACIER POINT: VIEW AND REVIEW

Start the Sierra Glacier Tour at Glacier Point on the south rim of Yosemite Valley, reached by automobile by the Glacier Point Road, which leaves Highway 41 about 5 miles south of Yosemite Valley. There are no campgrounds or cabins at Glacier Point, but there is camping at Bridalveil Creek about halfway between Highway 41 and Glacier Point.

Glacier Point was under 500 feet of ice at least once during the Sherwin glaciation about a million years ago. At that time Yosemite Valley was filled and ice overflowed the valley walls. Evidence for this are glacial erratics at various places between Glacier Point and nearby Sentinel Dome and at several places on the north side of the valley.

Glacier Point is a good place to review topics from other chapters. The flat floor of Yosemite Valley, 3,200 feet below, reminds us of Lake Yosemite, now in-filled with sediment. You can clearly see where John Muir's five glaciers once entered Yosemite Valley. Proceeding clockwise from the north they were: the Yosemite Creek Glacier, where Yosemite Falls is today; the Hoffman Glacier, just east of North Dome; the Tenaya Glacier, through Tenaya Canyon; the South Lyell Glacier, through Little Yosemite Valley; the Illilouette Glacier, along Illilouette Creek from the south, just east of Glacier Point. The Tenaya and the South Lyell (also called the Merced Glacier) were the

Figure 66. Yosemite Falls from Glacier Point. Above the Yosemite cliffs is an old landscape with gentle slopes that has been changed only slightly by glaciation. The many rounded granitic domes are the result of exfoliation, a weathering process, not glaciation.

largest of the five tributary glaciers and were the only two to reach the valley during the Tioga glaciation 20,000 years ago.

The steep, somewhat parallel cliffs and the flat valley floor suggest the "bottom dropped out" idea of Josiah Whitney. The valley is not a graben, but it does resemble one. Little Yosemite Valley was the location of one of the largest of the glacier pavements John Muir wrote about, shining "as if polished afresh every day." The glacial steps at Vernal Fall and Nevada Fall suggest the importance of joints to glacial erosion. The immense, glacially smoothed wall of Clouds Rest and the V-shape of the upper part of Tenaya Canyon are clearly visible. Despite the fact that large glaciers flowed through the canyon more than once, the ice had little effect here where the rock has few joints. The gentle land above the Yosemite cliffs (fig. 66) suggests the preglacial landscape Matthes re-created, parts of which remain where glaciers either did not reach or were too small to alter.

GLACIER POINT: HALF DOME

Half Dome (fig. 67), perhaps the most striking feature seen from Glacier Point, is one of Earth's unique rock sculptures. The great cliff coincides with a vertical joint, and rock has somehow been removed from one side of the

Figure 67. Half Dome from Glacier Point. At the time of the greatest glaciation, about 1 million years ago, glacial ice reached about halfway up the vertical cliff but never completely covered the mountain.

crack. Did the glaciers that flowed through Tenaya Canyon remove one side of the dome? Perhaps, but do not imagine ice flowing past the cliff grinding rock away, like sandpaper, until only what we see is left. Ice never covered Half Dome, but it did once fill the valley to within 700 feet of the summit. Possibly the flowing ice removed blocks of jointed rock one by one and, by undermining, caused the fall of joint-blocks from higher up the mountainside until the cliff was developed. It is also possible that rockfall triggered by an earthquake created the cliff in a few minutes during some interglacial time without any involvement by a glacier at all.

GLACIER POINT: HORNS

The peaks visible on the Sierra crest around Mount Lyell, 20 miles to the east, and the closer Mount Clark, are horns. Prior to glaciation the mountains were larger and rounded with few or no cliffs. They had the characteristic form of mountains shaped by weathering, rain, and river; imagine the Appalachian Mountains of today relocated to California. At various times during the Ice Age, glaciers formed on two or three sides of these rounded mountains and excavated cirques by headward erosion. As the cirques became larger the volume of rock that remained between them became progressively smaller until, when glaciation ended, all that remained of the former broad hill was a steep-cliffed pyramid or spire.

Figure 68. Bridalveil Meadow Moraine in roadcut near the Bridalveil Fall parking lot. The moraine is covered with trees and is difficult to see and photograph. The best way to appreciate it is to walk along the crest to see that it is indeed a ridge of till.

YOSEMITE VALLEY: BRIDALVEIL MEADOW MORAINE

From Glacier Point return to Highway 41, then go north to Yosemite Valley. Moraines of the Tioga-age Yosemite Glacier are present at the west end of Yosemite Valley in the vicinity of Bridalveil Meadow and El Capitan Meadow. Not many visitors notice these moraines, the best evidence of former glaciers in Yosemite Valley, because they are not as striking as the cliffs and waterfalls, their locations are not well marked, and they are largely overgrown with trees.

The Bridalveil Meadow Moraine (fig. 68) is the terminal moraine of the Yosemite Glacier of Tioga age and marks the western limit of the ice about 20,000 years ago. Earlier glaciers were larger and one extended 10 miles farther west nearly to El Portal. The moraine can be reached in two ways. Because traffic is often heavy and fast, study a road map of Yosemite Valley and the directions below before embarking on the quest.

First way to the moraine: Travel on the one-way road at the west end of the valley that takes traffic from Highway 140 on the north to Highway 41 on the south. As you proceed in the right lane, watch for a large dirt turnout on the right with a wood sign that says that President Theodore Roosevelt and John Muir once camped at this site. Park here. Walk along the road (in the direction of traffic) for about 100 yards, staying out of the meadow until you reach trees. Then walk into the meadow (left, or north), watching for a trail that goes northeast through trees skirting the meadow. It is a quarter mile to the Bridalveil Meadow Moraine, which is to the right, overgrown with trees. The trail takes you to a large erratic boulder, once part of the moraine, on which is a memorial to Dr. Lafayette Bunnell, the man who proposed the name "Yosemite." The moraine is in disrepair here because of erosion by the nearby Merced River, but to the right it is well preserved, showing as a tree-covered wall with boulders of all sizes protruding through a mantle of leaves and pine needles. Climb the 30 feet or so up to the crest of the moraine to see both sides and convince yourself that it is a ridge. Walk the crest of the moraine back to the road, then right against traffic back to your car.

Second way to the Bridalveil Meadow Moraine: Once again, start on the one-way road that goes southeast from Highway 140 to Highway 41, but this time drive in the left lane and go past the Roosevelt-Muir parking area described above. Watch for a sign that says "Yosemite Valley Destinations." Pull off to the left and park on the shoulder, walk toward the river (away from the road), and find a pleasant trail that goes northwest (down-river) on the side of the moraine. It is easy to scramble onto the crest of the moraine and confirm that it is indeed a ridge and not simply the left bank of the Merced River. Follow the crest to the Bunnell marker mentioned previously.

YOSEMITE VALLEY: "MEDIAL MORAINE"

A large moraine at the east end of the valley is called "Medial Moraine," even though its origin is in doubt. Eliot Blackwelder says it is a medial moraine formed where the left lateral moraine of the Tenaya Canyon Glacier merged with the right lateral moraine of the Merced River Glacier at the base of Half Dome. François Matthes, however, believes it is a recessional moraine of the Tenaya Canyon Glacier. I believe Blackwelder was correct.

Whether medial or recessional, it is a fine example of a moraine, is easy to get to, and is well worth a visit. By automobile or shuttle go to the horse stables at the east end of the valley (in 1996, stop 18 on the shuttle bus). The moraine is across the road (east of the road) from the stables. There are trails going east along both sides of the moraine, but to see it best walk the crest, about 60 feet above the river, for a third of a mile to a road cut. The moraine continues east of the road, up to the cliff at the base of Half Dome, but becomes covered by large boulders that have fallen from the mountainside above.

VERNAL FALL AND NEVADA FALL DAY HIKE

Take the shuttle bus to Happy Isles at the east end of Yosemite Valley to the trailhead to Vernal Fall and Nevada Fall, both of which are formed by the

Figure 69. The Merced River descends from Little Yosemite Valley to Yosemite Valley over two glacial steps, forming Nevada Fall and then Vernal Fall.

Merced River flowing over glacial steps (fig. 69). The hike to the top of Nevada Fall is a 6-mile round-trip and is moderately strenuous; for most people it will take almost all day to complete, but the scenery is good, and it is a good hike even if you only go part way. Consult the park booklet and map; a good plan is to go up the Mist Trail, cross the river above Nevada Fall, and return by the John Muir Trail south of Nevada Fall.

The Ice Age Merced Glacier originated miles from here in the vicinity of Mount Lyell on the Sierra crest. It flowed through Little Yosemite Valley and formed two glacial steps where it encountered vertically jointed rock near the entrance to Yosemite Valley. The ice plucked joint-blocks and carried them away. Because the joints were vertical, near-vertical cliffs resulted, and an icefall formed on the glacier giving it more erosional power. When the glacier melted there were two large cliffs in the path of the Merced River. The step at Vernal Fall is 317 feet high, and the step at Nevada Fall is 594 feet. Connecting Yosemite Valley with the higher Little Yosemite Valley, these glacial steps form a stairway of giant proportions, hence names such as giant's stairway and Cyclopean staircase are sometimes used to describe the scene.

At the top of Nevada Fall notice how little the river has cut into the granitic rock in about 15,000 years, the approximate length of time since the ice withdrew. The Merced River carries little sediment below Merced Lake

8 miles upstream and without sediment as an abrasive, running water cannot erode granitic rock effectively. There are well-preserved patches of glacial polish on the granitic rock at the top of Nevada Fall, south of the river, but beware of the cliff and keep youngsters on a leash.

MORAINE DOME BACKPACK

Moraine Dome is a backpack destination from the trailhead at Happy Isles at the east end of Yosemite Valley. Here you can see glacial polish, glacial erratics, a moraine, and evidence of multiple glaciation. Jeffrey Schaffer's hiking guide *Yosemite National Park* is recommended, along with U.S.G.S. 7.5′ Half Dome and Merced Peak topographic quadrangles. The hike is about 14 miles round-trip from Happy Isles and will take three days. Expect bears.

From the top of Nevada Fall (preceding section) go northeast and east on the John Muir Trail into Little Yosemite Valley, so named because it is relatively wide and has a flat floor somewhat resembling Yosemite Valley. Along the Merced River between Yosemite Valley and Merced Lake there are alternating wide and narrow stretches of valley. The names Little Yosemite Valley, Lost Valley, and Echo Valley are associated with wide places. Joints are the cause of the variation in width along the path of the large Merced Glacier. Where the rock is highly jointed the glacier eroded effectively and the valley is wide; where joints are lacking or widely spaced the valley is narrower.

About a mile east of Nevada Fall the John Muir Trail goes north along Sunrise Creek, away from the Merced River. Moraines are present on the north side of Little Yosemite Valley; look for ridges 20 or 30 feet high composed of sand, gravel, and boulders all mixed together, though often all you see initially is a hillside with trees, pine needles, and half-exposed boulders. If you climb up the "hillside" and find you are on a ridge with boulders and slopes on two sides, it is a moraine.

In the next 2 miles avoid taking either the Half Dome Trail or the Clouds Rest Trail, which branch left from the John Muir Trail. A mile east of the Clouds Rest Trail junction our trail crosses Sunrise Creek. Study the map and see that Moraine Dome is only a quarter of a mile southeast of here. The southeast side of the dome is steep and unsafe. A ridge connects Moraine Dome with a second, unnamed, dome a half mile to the northeast.

Moraine Dome was named for a prominent moraine on its south side (fig. 70), one of many moraines on the north slope of Little Yosemite Valley. The moraine on Moraine Dome is 150 feet below the summit and 1,800 feet above the floor of Little Yosemite Valley, which indicates the thickness of the glacier. Detailed mapping shows that this moraine, which can be traced

Figure 70. Tioga moraine on Moraine Dome. (Photo by F. E. Matthes, U.S. Geological Survey.)

north to Sunrise Mountain, is the highest moraine of the Tioga glaciation of about 20,000 years ago.

But there is more. On Moraine Dome, above the Tioga moraine, there is evidence that ice covered the dome previously. A prominent boulder 12 feet long, a glacial erratic (fig. 71), perches on a pedestal of rock 3 feet higher than surrounding granitic rock of the dome. That this boulder was deposited by ice is shown by the fact that it consists of a different type of granitic rock than that of which Moraine Dome is made. A glacier deposited this boulder on Moraine Dome so long ago that weathering since then has lowered the surrounding surface 3 feet (or more) except where the erratic provided protection. Another erratic boulder is also present on top of the dome, fallen from a nearby pedestal. These two erratic boulders are all that remain of a moraine that once covered the summit of Moraine Dome. But the Tioga moraine below the summit on a steep slope is well preserved and shows little effect from perhaps 15,000 years of exposure. It seems clear that the higher moraine, now mostly destroyed, was very old when the Tioga glacier arrived.

There is another interesting feature on Moraine Dome. Close to the summit is a wall of rock 15 feet long, 7 feet high, and 4 feet thick (fig. 72). The rock is aplite, a special sort of granitic rock that forms in long narrow masses that some people call veins, but which are properly called dikes. Such dikes vary from an inch to several feet thick and may be hundreds of feet long. Aplite dikes are common in the granitic rocks of the Sierra Nevada and are usually more durable than the rock into which they were intruded; com-

Figure 71. Erratic boulder on rock pedestal at Moraine Dome. The boulder protected the rock underneath from weathering during the approximately 1 million years since the time of the Sherwin glaciation. There may have been other pedestals from which the boulder fell before this one formed, so the three-foot height represents the minimum amount the dome has been lowered since glaciation. (Photo by G. K. Gilbert, shown in the photo, U.S. Geological Survey.)

monly they preserve glacial polish long after the surrounding rock has weathered away, and often they can be found etched into relief by weathering. The rock wall on Moraine Dome is part of a vertical aplite dike that, because of weathering, now extends 7 feet above the surface of the granitic rock it invaded.

The rock wall on Moraine Dome is so fragile it could not have survived the glaciation that left the erratic boulders. Thus it seems clear that ice of a pre-Tioga glaciation flowed over Moraine Dome, smoothed and polished both the granitic rock and the aplite dike to the same level, deposited a moraine, then withdrew. Then during many millennia of exposure, weathering and erosion removed at least 7 feet of rock from the dome and destroyed the moraine except for two boulders. At many other places in the Sierra we see where glaciers have smoothed granitic rock and aplite dikes to a common level, and approximately 15,000 years of weathering has produced only a fraction of an inch of relief between the two rock types. How many years, then, has it taken to produce the aplite wall and the rock pedestals at Moraine Dome? Between 750,000 and 1,000,000 years, the length of time since the Sherwin glaciation.

Figure 72. A vertical dike of aplite on Moraine Dome protrudes 7 feet above the surface of the granitic rock. The tough aplite has weathered into relief in the approximately 1 million years since it was eroded flush with the granitic rock during the Sherwin glaciation. Compare with figures 18 and 78. (Photo by G. K. Gilbert, U.S. Geological Survey.)

It is instructive and valuable to see things like the erratic boulders and the rock wall for oneself and to ponder the question, "If I had visited Moraine Dome before François Matthes did in 1906, would I have seen and understood?" So few people have the opportunity, or make the effort, to observe natural phenomena at all. Of the few who do, only a small number are able to decipher the clues present in the mute rock. Visiting places like Moraine Dome helps us appreciate the accomplishments of persons like King, Russell, Muir, Matthes, and others, who have managed to see, understand, and tell others.

SIESTA LAKE: TILL IN ROADCUT

From Yosemite Valley go west on Highway 140, and then northwest on the Big Oak Flat Road to Highway 120, then east on the Tioga Road toward Tuolumne Meadows and Tioga Pass. There are some campgrounds along this route for which reservations are not required.

About 10 miles along the Tioga Road is small Siesta Lake, with parking just past the lake on the right. An exhibit explains how the lake was formed by a moraine damming a creek, but this condition is not readily visible. What can be seen is a good exposure of till in the roadcut, permitting close examination of the material of which moraines are built. Boulders, gravel, and sand are all mixed together without any significant layering. Lack of sorting and layering is a diagnostic characteristic of till and makes it easy to distinguish from river deposits. Because of its density and surficial solidity, ice can carry large boulders and sand grains side-by-side, at the same speed, and deposit them at the same place. While a river in flood can also move boulders along with sand and mud, as the water slows down it will first deposit the coarser material, then at a slower speed it will deposit sand, and then, only at the slowest speed, it will deposit mud. The result is that materials of different sizes are deposited at different times and places, sorted by size, and layered, quite unlike till.

MOUNT HOFFMAN DAY HIKE

About 3 miles east of the Porcupine Flat Campground is the road to May Lake High Sierra Camp and the trailhead for a day hike to the top of Mount Hoffman. Meals and lodging are possible at the High Sierra Camp by reservation.

The climb to the 10,850-foot summit of Mount Hoffman is strenuous but not dangerous. Hundreds, perhaps thousands, of people climb it each summer because it yields a high return for the time and energy invested. If you have never climbed a High Sierra peak before, this is the one to start with. Take lunch, water, and warm clothing, leave early, and turn back if thunderstorms develop. The hike is 6 miles round-trip; plan 6 or 7 hours total. High on the mountain, the trail is sometimes indistinct, but if you simply keep going up and avoid steep cliffs you will be on the correct route.

There are many things to see, but our interest is in the difference in landscape south and north of the summit. Prior to glaciation Mount Hoffman was a somewhat rounded hill, part of a subdued, gentle landscape formed by weathering and river erosion. The land south of the summit still has much of this character. From Glacier Point this old landscape was visible above the cliffs of Yosemite Valley, and you have been traveling over it along the Tioga Road. However, north from the summit of Mount Hoffman are steep slopes and cliffs characteristic of a glaciated landscape. A glacier has excavated a bold cirque into the north slope of Mount Hoffman, and you are at the top of its steep headwall, with a cirque lake below. Headward erosion by glaciers excavated cirques, and other glacial erosion created the topography visible to the north.

MORAINE AT MARKER T22

A few miles farther along is road marker T22, where the road cuts through a large moraine and produces a good exposure of till (see Siesta Lake for discussion). There is good parking just past the roadcut.

OLMSTED POINT

Olmsted Point, a prominent turnout (to the right) about halfway between the May Lake turnoff and Tenaya Lake offers good views, glacial polish, and glacial erratics. A nature trail leaves the parking area off to the right and turns back to below the parking area. Along the way are numerous erratic boulders and patches of glacial polish on the granitic rock, though little of it shines "as if polished afresh every day"; perhaps the custodian has been inattentive to duty.

TENAYA LAKE

Two miles east of Olmsted Point is Tenaya Lake. The Tenaya Basin was eroded by ice from the highland ice field in Tuolumne Meadows, which overflowed a low divide to Tenaya Lake and then traveled down Tenaya Canyon into Yosemite Valley; this tongue of ice was one of the two main tributaries of the Yosemite Glacier. Glaciers widened the preglacial valley and excavated the large, deep basin occupied by Tenaya Lake. Three miles south, however, on its way to Yosemite Valley, Tenaya Creek enters narrow Tenaya Canyon. The narrowness is due to a sparsity of joints in the granitic rock, which prevented the glacier from widening Tenaya Canyon. This is why there is no trail along Tenaya Creek to Yosemite Valley, and why a note on the park map states, "Hiking in Tenaya Canyon is dangerous and not recommended."

To see one of Muir's glacial pavements with large areas of well-preserved polished rock, stop at the southwest end of the lake (or take the shuttle bus from Tuolumne Meadows) and explore the area across the road from the lake, walking northeast (figs. 73, 74). Everywhere there are erratic boulders and large patches of polished granitic rock that give brilliant reflections when the sun is right. Look for aplite dikes (see Moraine Dome, this chapter) with polish and striae, which have been destroyed on adjacent, coarser granitic rock. Some of the aplite has been etched into relief by weathering in the 15,000 years or so of exposure since the ice withdrew.

Tenaya Lake is 114 feet deep, attributable in part to glacial excavation and in part to damming by till. At the northeast end of the lake is a picnic area on a delta that Tenaya Creek is building into the lake. Unless the mud, sand,

Figure 73. Glacial polish shines "as if polished afresh every day" (John Muir) on a glacier pavement at Tenaya Lake, Sierra Nevada. Why is the polish weathering away in circles? The dark objects are small pine cones washed down from above.

Figure 74. Large erratic boulder on a glacier pavement at Tenaya Lake, Sierra Nevada.

and gravel deposited by the creek is cleaned out during some future glaciation, Tenaya Lake eventually will be filled with sediment and become a meadow, with Tenaya Creek meandering through it, as has already happened to many smaller Sierra lakes.

All of the Tenaya Basin was eroded and abraded by ice, which was 2,460 feet thick and exerted a pressure of about 74 tons per square foot at the level of the present lake. Polly Dome north of the lake and Tenaya Peak southeast of the lake were overridden by ice, and in many places glacier-polished rock reflects sunlight from high on the mountainsides. Roches moutonnées of all sizes are present, including large Pywiack Dome northeast of the lake.

TUOLUMNE MEADOWS

Two and a half miles past the northeast end of Tenaya Lake the Tioga Road crosses an inconspicuous divide between Merced River and Tuolumne River drainages, and soon thereafter enters Tuolumne Meadows. Here there is a store, a small restaurant, a large campground, and tent cabins with meals (reservations probably needed).

Two maps by the U.S. Geological Survey are nice to have when visiting Tuolumne Meadows. One is a geologic map of the area (Bateman, Kistler, Peck, and Busacca, 1983), and the other shows the distribution of glaciers 20,000 years ago (Alpha, Wahrhaftig, and Huber, 1987). Tuolumne Meadows is where Ice Age glaciation was first recognized in California, in 1863. A large highland ice field covered all of the area south to Mount Lyell, and it was here that the 60-mile-long Tuolumne Glacier, longest ever in California, originated. Ice about 2,000 feet thick covered Tuolumne Meadows, as revealed by the height of the boundary between smooth glaciated rock and rough rock on Unicorn and Cathedral Peaks south of the highway. Glaciation has removed lots of rock from the Tuolumne Meadows area, leaving a wide, open landscape surrounded by high residual peaks.

As the last Ice Age glacier stagnated and melted, perhaps 15,000 years ago, a blanket of till and glacial outwash was deposited over most of Tuolumne Meadows. The resulting land surface was rather flat and had poor drainage, so many marshes, ponds, and shallow lakes were present. Some of these have been in-filled to form meadows, but many marshes and ponds remain, and the shallowness of the water table keeps the meadows open by making it difficult for trees to grow.

POTHOLE DOME DAY HIKE

At the west end of Tuolumne Meadows is Pothole Dome, reached by auto or by shuttle bus from the store near the campground. It is a large roche mou-

Figure 75. Glacial polish and potholes on the south side of Pothole Dome, Tuolumne Meadows, Sierra Nevada. The potholes formed in the bed of a river that flowed under the ice when the Tuolumne Glacier covered this area about 20,000 years ago.

tonnée with a smoothed, gentle east slope (faced upstream) and a steep, ragged west slope (faced downstream) fashioned by the Tuolumne Glacier. You can easily walk to the dome from a turnout on the highway and see, on the south side of the dome along the trail, potholes several feet in diameter. Also watch for erratic boulders, aplite dikes (see Moraine Dome, this chapter), and large areas of glacial polish that are quite slippery (figs. 75–78). The climb to the summit of the dome is highly recommended and is easy up the gentle slope from the east end. Many interesting glacial features can be seen by walking north from the summit and returning along the base of the dome.

Potholes are the unusual attraction here. Potholes are circular holes with steep sides excavated into a riverbed by cobbles or boulders being spun around continuously in a circular turbulent eddy. The spinning rocks, called grinders, drill holes into the rock of the streambed by abrasion. In so doing, the grinders are worn out: cobbles and boulders end up as sand grains. Potholes vary in depth and diameter from a few inches up to 10 feet or so and are common in channels of turbulent mountain streams or rivers. Stepping into a pothole and being trapped by the swirling water is a possibly fatal hazard for persons swimming in mountain streams.

But the potholes of Pothole Dome are on the side of a hill, not in a river channel. The explanation is that they formed in a river that flowed under the Tuolumne Glacier, which was over a thousand feet thick here. The river was

Figure 76. A pothole at Pothole Dome with remains of boulders that did the drilling. Rocks trapped in turbulent circular eddies whirl around and around, and drill potholes by abrasion.

Figure 77. Erratic boulders on a glacier pavement at Pothole Dome, Tuolumne Meadows.

Figure 78. Glacial polish remains partially intact on an aplite dike even though surrounding granitic rock has been lowered about an inch by weathering in the approximate 15,000 years since it emerged from under the ice. Photographed at Pothole Dome, Tuolumne Meadows. Compare with figures 18 and 72.

fed by meltwater that flowed into moulins on the surface of the glacier; glacier ice formed the walls of the river channel, and rock formed the bottom, where the potholes developed.

LEMBERT DOME DAY HIKE

Lembert Dome is a prominent landmark near the store and campground at Tuolumne Meadows. It is about 500 feet high and, like Pothole Dome, has the characteristic shape of a roche moutonnée. Ice 1,500 feet thick covered the summit during the Ice Age. Rock around the base of the dome, near the parking lot, has good glacial polish, as does rock elsewhere on the dome. It is not difficult to climb to the summit if you go east from the campground along the highway for half or three-quarters of a mile, then up the gentle slope; if the slope seems too steep, go farther east. Watch for erratic boulders in addition to slippery glacial polish.

LYELL AND MACLURE GLACIERS BACKPACK

Glaciers on Mounts Lyell and Maclure can be visited by a three-day backpack trip from Tuolumne Meadows. Consult the Vogelsang Peak and Mount Lyell 7.5′ topographic quadrangles, and Hike 58 of Schaffer (1992).

Follow the John Muir Trail southeast up the gentle slope of the beautiful, U-shaped, glacial trough occupied by the Lyell Fork of the Tuolumne River for 8 or 9 miles to a good camping area. Be prepared for visits by bears. The glaciers are about 3 miles away, close together, in cirques at the head of the glacial trough. From camp go 1 or 2 miles south on the John Muir Trail before cutting cross-country, watching for small stacks of stones ("ducks") that may help you stay on the best route. The Lyell Glacier is divided into eastern and western parts by a ridge of rock. Watch for bare ice with crevasses, glacier tables, and bergschrunds. Mount Lyell, 13,114 feet, is usually climbed over the west part of the glacier.

MORAINES

Moraine Flat, 3 miles northeast of Tuolumne Meadows, is a good place to see a swarm of moraines and can be visited by persons with a topographic map and cross-country hiking skill; a copy of the geologic map of Tuolumne Meadows will be helpful (Bateman, Kistler, Peck, and Busacca, 1983). Between Tuolumne Meadows and Tioga Pass the road crosses several moraines, the largest of which is exposed in a roadcut at marker T33.

GAYLOR AND GRANITE LAKES DAY HIKE

Tioga Pass is 6 miles beyond Tuolumne Meadows at the east entrance to Yosemite National Park. By a moderate day hike you can see glacial lakes, kettle lakes, and a cirque. There are no campgrounds or accommodations at Tioga Pass but there are some just a few miles east. A parking lot just inside the park, less than a hundred yards from the entrance station, is the trailhead for a day hike to Gaylor and Granite Lakes. The hike of 4½ miles round-trip is only strenuous for the first half mile. Consult the Tioga Pass 7.5′ topographic quadrangle.

From the parking lot the trail immediately starts climbing over a ridge; persevere, this is the hard part. On the way up the trail crosses a small lateral moraine. After a half mile the top of the ridge yields fine views east and west. To the east is Highway 120 and Dana Meadows. The numerous small lakes and ponds in the meadows (fig. 79) are in glacial deposits left by the last glacier and are probably kettle lakes. Kettles form as follows. Masses of ice, isolated and abandoned by a retreating glacier, are partially or completely buried under till and glacier outwash. After the ice melts a depression remains in the sediment approximating the size and shape of the original mass of ice. If the kettle holds water it becomes a kettle lake.

Looking west, Middle Gaylor Lake is at the bottom of the ridge, and there is a good view of the high peaks of the Cathedral Range south of Tuolumne

Figure 79. Mount Dana and Dana Meadows near Tioga Pass, Sierra Nevada. The ponds and small lakes are in kettles, shallow depressions formed where isolated pieces of a wasting glacier became buried or surrounded by sediment and later melted.

Meadows. To the northwest is a large cirque (fig. 80) occupied by Upper Granite Lake. Descend to the inlet of Middle Gaylor Lake and follow the trail up along the small stream toward Upper Gaylor Lake. Leave the trail anywhere before Upper Gaylor Lake and walk cross-country for a quarter mile, west or northwest, across a low, broad ridge covered with till and erratic boulders, to beautiful Lower Granite Lake. Follow the east shore of Lower Granite Lake to Upper Granite Lake for good views of the cirque at the head of the valley. The highly jointed granitic rock that is prominent in the headwall facilitated excavation of this wide cirque. Recall W. D. Johnson's descent into the bergschrund of the Lyell Glacier and his description of blocks of rock in various stages of being pulled from the headwall by the ice; by such plucking a glacier erodes headward and a cirque such as this is excavated.

From the Granite Lakes you can see the full extent of the Cathedral Range south to Mounts Lyell and Maclure and their glaciers far in the distance.

Figure 80. Middle Gaylor Lake and the cirque where Upper Granite Lake is situated, Sierra Nevada. This cirque is wider and less deep than many cirques are. The low ridge in the middle distance is covered with till.

DANA GLACIER DAY HIKE

Tioga Lake, a mile past Tioga Pass on the highway, has a campground on the west side of the lake. From here you can visit the Dana Glacier. The Tioga Pass and Mount Dana 7.5′ topographic quadrangles show the area. The hike to the glacier is about 5 miles round-trip, with 2,000 feet of climbing to the terminus of the glacier; plan a 6-hour round-trip, plus whatever time you spend at the glacier. There is no official trail but the route is clear; follow the stream up Glacier Canyon past several lakes. Dana Glacier lies at 11,600 feet elevation in a cirque on the north face of Mount Dana. A large couloir extends from the glacier to a ridge near the summit. I am told that in summer some expert skiers climb the mountain and ski down the couloir and glacier. I do not recommend this.

NUNATAK NATURE TRAIL

Just east of Tioga Lake, on the left side of the road, is a large parking area for the Nunatak Nature Trail, a quarter-mile paved trail that is highly recommended. The ice field in this part of the Sierra covered everything except the highest peaks and ridges, which stuck out as islands of rock above a sea of ice; mountaintops in this predicament are called nunataks. Excellent exhibits focus on the Ice Age and how nunataks helped plants and animals survive and reclaim the barren land exposed as the glaciers melted. Living on a nunatak could not have been easy, but it was possible, as the exhibits explain.

CONNESS GLACIER VIEW DAY HIKE

From Tioga Lake go northeast on the highway about a mile to Saddlebag Lake Road, and then go left (north) 3 miles over alternating pavement and gravel to Saddlebag Lake. There is a store and a National Forest Service campground at the lake. From here a boat ride and short hike permit you to see cirques, glacial erratics, glacial polish, a glacial step, and the Conness Glacier with its bergschrund. Bring binoculars and the Tioga Pass 7.5′ topographic quadrangle.

Get to the northwest end of Saddlebag Lake either by walking around the west side of the lake or by taking a water taxi from the store ($6 round-trip in 1996). The round-trip hike to Steelhead Lake from the north end of Saddlebag Lake is an easy 3 miles. From the boat dock at the north end of Saddlebag Lake there is a good view south to Mount Dana, its glacier, and its couloir. Walk northwest along a wide trail, a former mining road, past Greenstone Lake toward Steelhead Lake. Around the east shore of Greenstone Lake, and all along the route to come, are erratic boulders of light-colored granitic rock resting on dark-colored rock of metamorphic origin (fig. 81). These erratics were carried by ice from their source in granitic rocks, visible to the west, and deposited here when the last Ice Age glacier (Tioga) underwent its final retreat, perhaps 15,000 years ago. The contrasting colors of erratics and bedrock make the erratics conspicuous and prove the boulders could not have formed in place by weathering. Some erratics near Steelhead Lake are quite large and are in precarious positions above the trail. Glacially polished and rounded rock outcrops are common all along the trail.

After about three-quarters of a mile a small stream from Z Lake crosses the trail. From here the Conness Glacier is visible to the southwest partially filling a large cirque on the northeast slope of Mount Conness, elevation 12,590 feet. A glacial step with a waterfall and cascade is below the glacier. With binoculars you can see moraines at the base of the glacier covering the ice, and usually a bergschrund is visible as a line, or series of discontinuous lines, parallel to the mountainside just below the contact of ice and rock at the head of the glacier. Experienced hikers or climbers can visit the Conness Glacier by following an informal trail along the inlet creek at the west end of Greenstone Lake. The trail ascends the glacial step in the brush to the right of the waterfall.

Figure 81. A glacial erratic of light-colored granitic rock rests on dark metamorphic rock near Steelhead Lake, Sierra Nevada, where a melting glacier left it about 15,000 years ago. The presence of different types of rock proves the boulder has been transported.

Figure 82. Glacieret on North Peak, Sierra Nevada, July 1996. Meltwater is seeping through the moraine. Moraines of some glaciers hold meltwater, suggesting that they have an impermeable core of ice.

Continue north to Steelhead Lake for even better views of the Conness Glacier and a good view of a glacieret on the northeast side of North Peak (fig. 82). A bergschrund is sometimes visible at the head of this small ice body. Lovely Cascade Lake, a quarter mile west of Steelhead Lake, is named for the cascades of meltwater issuing from the glacieret. The lake can be visited by an unofficial trail that leaves the main trail a few hundred yards south of Steelhead Lake and goes west over a low ridge.

MORAINES OF LEE VINING CANYON

From Saddlebag Lake, return to Highway 120, then travel east toward its junction with Highway 395 and the town of Lee Vining. The highway follows the north side of Lee Vining Canyon, providing thrills and white knuckles for many people because of the steep drop-off. There are several large pullouts for viewing and photography. We will visit moraines in the lower part of the canyon.

Near the bottom of the steep grade are several campgrounds with signs. Turn off the highway to the right (south) at the first campground sign and follow the sign to Moraine Campground. Park anywhere in the campground, much of which is on a series of recessional moraines that cross the canyon between lateral moraines that are over 700 feet high (fig. 83). The recessional moraines fill the valley except where Lee Vining Creek and the highway have cut through them. To inspect the crests of different moraines, walk for about a quarter mile down-valley.

Return to the highway and continue east for a half mile, and then turn off to the right (south) into the Cattleguard Campground. Go to campsite 5, which is on the large mass of till formed by the several recessional moraines. A hundred feet east of here you can look down the steep, east-facing slope of the moraine into another, lower campground. From here is a good view of the lateral moraines, the recessional moraine, and Lee Vining Creek flowing noisily a hundred feet below, perhaps still cutting down through the moraine that crosses its path.

Drive back toward the highway, but before reaching it turn right to campsites 1 and 2 and park at the turnaround. You are once again on the crest of a recessional moraine (fig. 84). Walk north a hundred feet until you can look east into the meadow and see yet another recessional moraine, rather small, extending across the meadow perpendicular to the length of the valley.

Return to Highway 120 and turn off, again to the right, at the next road for a good view back to the large recessional moraine that you were just on, as well as numerous avalanche chutes and cirques at the head of the canyon (fig. 85).

Figure 83. The long ridge is a right-lateral moraine in Lee Vining Canyon. Lateral moraines in the Sierra Nevada are often much larger than many people expect.

Figure 84. The crest of a recessional moraine in Lee Vining Canyon extends across the valley perpendicular to the lateral moraine in the distance.

The guided trip ends here, but many interesting canyons can be explored by side roads and walking trails off Highway 395, both north and south of here. Some of these places have been mentioned and illustrated in previous chapters. Almost every canyon has moraines, glacial lakes, and other glacial features that can be traced, pleasantly, with camera and fly rod, up the glacial valleys to their sources in the silent cirques of the High Sierra.

Figure 85. A recessional moraine extends across Lee Vining Canyon from left to right, just above the truck. The lower ends of avalanche chutes high on the mountain mark the elevation of a glacier that covered this area thousands of years ago. Rounded forms predominate below the former level of the ice.

REFERENCES CITED

Alpha, Tau Rho, Clyde Wahrhaftig, and N. King Huber. 1987. *Oblique map showing maximum extent of 20,000-year-old (Tioga) glaciers, Yosemite National Park, California.* U.S. Geol. Survey Misc. Investigations Series Map I-1885.

Bateman, P. C., R. W. Kistler, D. L. Peck, and A. J. Busacca. 1983. *Geologic map of the Tuolumne Meadows quadrangle, Yosemite National Park, California.* U.S. Geol. Survey Geologic Quadrangle Map GQ-1570, scale 1:62,500.

Ferguson, Sue A. 1992. *Glaciers of North America: A field guide.* Golden, Colo.: Fulcrum Publishing.

Schaffer, Jeffrey P. 1992. *Yosemite National Park.* 3rd ed. Berkeley: Wilderness Press. 1994 update.

Trent, D. D. 1996. Personal communication.

GLOSSARY

Ablation Loss of snow or ice from a glacier by melting, evaporation, or blowing away.

Alpine glacier *See* Mountain glacier.

Arête A narrow ridge between horns produced by glacial erosion.

Avalanche chute A smooth, steep, funnel-shaped indentation into a mountainside eroded by repeated avalanches of snow.

Basal sapping The process whereby a mountain glacier extends itself headward into a mountainside by plucking, creating a cirque.

Bergschrund A crevasse at the head of a mountain glacier formed where ice thick enough to flow pulls away from stationary ice and snow.

Biscuit-board topography A landscape with lots of semicircular excavations (cirques) produced by headward erosion of valley glaciers into an older, gently sloped landscape, remnants of which remain between the cirques. The cirques are what the "biscuit-cutter" removed.

Cirque An excavation into a mountainside at the head of a glacial valley produced by headward erosion of a glacier. Cirques commonly take the shape of a three-sided bowl and have steep cliffs. Rhymes with "jerk."

Cirque glacier A small glacier located in, and restricted to, a cirque. Cirque glaciers are often wider than they are long.

Cirque lake A lake in a cirque formed by a dam of bedrock and/or a moraine.

Col — A low spot, saddle, or pass on an arête. Rhymes with "doll."

Continental glacier — A very large ice sheet covering several million square miles.

Couloir — A steep valley or gully at the head of a glacier, filled with ice and snow, that extends from the glacier up to a ridge. A couloir often provides a climbing route from a glacier to the summit of a mountain. Rhymes with "cool car."

Crag and tail — *See* Roche moutonnée.

Crevasse — An open crack, or fracture, in the ice of a glacier.

Cyclopean staircase — *See* Giant's staircase.

Drift — Sediment deposited by a glacier (till) or by meltwater issuing from a glacier (glacial outwash).

Erratic — A rock or boulder transported some distance by a glacier and left isolated when the glacier melted. Some erratics weigh many tons and are as large as small houses.

Firn — Snow more than a year old, in an intermediate stage between snow and ice. Rhymes with "burn."

Giant's staircase — Two or more glacial steps in a glacial valley. These form stairs "suitable for giants."

Glacial drift — *See* Drift.

Glacial episode — *See* Glaciation.

Glacial erratic — *See* Erratic.

Glacial flour — Mud in streams from a glacier. When it gives a white appearance to water the result is spoken of as *glacier milk.*

Glacial geology — Study of the effects of glaciers on the land and the history of glacial events.

Glacial lake — A lake formed by a glacier, either as the glacier excavated a depression in bedrock or deposited till that dammed a stream, or a combination of the two.

Glacial outwash — Sediment transported and deposited by meltwater issuing from a glacier. Differs from till in that it is somewhat layered and sorted into mud, sand, and gravel.

Glacial polish — Smooth, mirrorlike surface on rock that results from abrasion by mud-sized particles in ice at the base of a glacier.

Glacial quarrying — *See* Plucking.

Glacial stairway *See* Giant's staircase.

Glacial trough *See* Glacial valley.

Glacial valley A river valley modified by a glacier. When a valley glacier flows down a river valley, it usually widens and straightens the valley and changes it from a V-shaped to a U-shaped profile.

Glaciation A time of major growth of glaciers. The Ice Age consisted of numerous glaciations separated by interglaciations.

Glacier A mass of ice and snow that forms on land, lasts for many years, and is of sufficient thickness to flow downhill because of its own weight.

Glacieret 1. A small glacier. 2. A mass of snow and ice that is difficult to classify; it may be either a perennial snowbank or a small glacier.

Glacier milk *See* Glacial flour.

Glacier mill *See* Moulin.

Glaciology Study of glaciers and ice.

Grooves *See* Striations.

Hanging valley A tributary valley that joins the main valley at an elevation above the floor of the main valley. Hanging valleys form where a small tributary glacier cannot deepen its valley as fast as the main glacier can. Often there is a waterfall after the glacier is gone.

Highland ice field In mountains, an extensive area of ice formed by valley glaciers overfilling their valleys. Only the highest peaks rise above the surface of the ice. Many early California writers refer to these areas as ice caps.

Holocene Epoch The last 10,000 years of geologic time, after the end of the Pleistocene Epoch.

Horn A three-sided peak, or spire of rock, left behind after erosion of cirques has mostly consumed a once-larger mountain.

Ice Age The period of time from about 1.6 million years ago until 10,000 years ago, during which glaciers several times covered areas that today have no glaciers or only small glaciers. There have been ice ages throughout geologic time, but this most recent one is the best known, and it is understood that this is the one being referred to unless another is specifically designated by some other name.

Ice cap — *See* Ice sheet.

Ice field — *See* Highland ice field.

Icefall — The point where a glacier flows over a cliff or down a very steep slope. The glacier is cracked into blocks of irregular size and shape, which, by down-dropping or toppling over, descend the slope.

Ice sheet — A large unconfined mass of ice that flows in several directions from a center. An ice sheet has a surface area greater than 50,000 square kilometers, while an ice cap is smaller. Both have a convex-up surface with gentle slopes and may completely bury mountain ranges.

Interglaciation — A time of relative warmth between two major glacial advances. The Ice Age contained numerous glaciations separated by interglaciations.

Jökulhlaup — A flood produced by a glacier, perhaps caused by melting during a volcanic eruption or by a break in an ice dam where a glacier temporarily blocked a river. Pronounced "yo-kool-awp."

Kettle — A depression in sediment deposited by a glacier, formed where a mass of buried ice melted.

Kettle lake — A lake in a kettle.

Little Ice Age — A time of colder climate and renewed glacier advance during the last few hundred years, from about A.D. 1250 to 1900. *See* Neoglaciation.

Matterhorn peak — *See* Horn.

Moraine — A mass of mostly unsorted rock debris (till) in or on a glacier or deposited by a former glacier. Moraines often are in the form of ridges. A lateral moraine lies on the side of a glacier. An end moraine extends around the terminus of a glacier and connects the laterals. A terminal moraine is an end moraine at a glacier's most advanced position. Recessional moraines are a series of end moraines deposited by a receding glacier, which are progressively younger up-valley. A medial moraine forms where two glaciers join together, their lateral moraines merging in the middle of the new glacier. A ground moraine is an irregular, thin blanket of till deposited under the ice.

Moulin — A vertical hole or shaft in a glacier down which meltwater flows.

Mountain glacier A glacier that forms in, and is confined to, mountains.

Neoglacial *See* Neoglaciation.

Neoglaciation New glaciation. Glacial advances during the last 10,000 years, since the end of the Ice Age.

Névé 1. A synonym for firn. 2. The area, or place, where snow is converted into firn and ice, and glaciers originate. Almost rhymes with "bevy."

Nunatak An Inuit word meaning "lonely peak," referring to a hill or mountaintop completely surrounded by glacier ice; an island of rock in a sea of ice.

Outburst flood *See* Jökulhlaup.

Paternoster lakes Several lakes connected by a stream in a glacial valley.

Pleistocene Epoch A subdivision of geologic time largely coincident with the Ice Age. It began 1.6 million years ago and ended 10,000 years ago.

Pleistocene Ice Age *See* Ice Age.

Plucking Pulling apart of rock, usually along joints, by ice flowing away from a hillside and carrying pieces of rock with it. Plucking is an important process of glacial erosion.

Pluvial lake A lake that existed because of more rainfall and/or less evaporation than at present. There were many pluvial lakes in the Great Basin east of the Sierra Nevada during the Ice Age.

Protalus rampart A pile or ridge of rock fragments formed where rock falls from a cliff onto a snowbank, then slides or rolls across the snow and accumulates along the margin. When the snowbank melts, it leaves a pile of debris that resembles a moraine.

Quaternary period The last 1.6 million years of geologic time, including the Pleistocene and Holocene Epochs. Geologists, biologists, climatologists, archeologists, and others who study the many aspects of the Ice Age and more recent time are often called Quaternary scientists.

Roche moutonnée A rock or hill of distinctive shape sculptured by flowing ice. There is a gently sloping, smooth upstream side and a steep, rough downstream side. The shape results because rock is tough and durable when pushed by ice, but weak and easily pulled apart by ice flowing away from a hill or cliff. Pronounced "row-sh-moot-o-nay."

Rock glacier A tongue-shaped mass of boulders and smaller rock debris (talus) that has fallen from cliffs above and that shows evidence of having flowed downslope. There may or may not be an ice glacier under the rock debris.

Sapping *See* Plucking.

Serac A pinnacle, or tower, of ice at an icefall on a glacier. Rhymes with "back."

Sheep rock *See* Roche moutonnée.

Snow chute *See* Avalanche chute.

Striations Scratches or elongated grooves on a glaciated rock surface where a grain of sand or rock imbedded in the ice was dragged over the surface.

Tarn A small lake created by glacier erosion or deposition.

Till Rock debris deposited directly by a glacier. Clay, sand, gravel, and boulders are all mixed together without sorting or layering. Till is what moraines are mostly made of.

Trimline A line, or boundary, on a valley wall that marks the highest extent of a former glacier. A sharp trimline, with all vegetation removed below the level of the ice, indicates recent glacier advance and retreat. If there is no vegetation, it is evident that rock that was below the level of the ice looks different from that above: rock below is smooth and rounded, while rock above is rough and craggy.

Valley glacier A long mountain glacier flowing like a river between valley walls.

ADDITIONAL READING

Note: Copies of most of the references used in preparing this book are conveniently located in a single file at the Special Collections Desk, Meriam Library, California State University, Chico. Here are some additional sources of related information.

Farquhar, Francis P. 1966. *History of the Sierra Nevada.* Berkeley: Univ. of Calif. Press.

Ferguson, Sue A. 1992. *Glaciers of North America: A field guide.* Golden, Colo.: Fulcrum Publishing.

Fryxell, Fritiof, ed. 1962. *François Matthes and the marks of time.* San Francisco: Sierra Club.

Hill, Mary. 1975. *Geology of the Sierra Nevada.* Berkeley: Univ. of Calif. Press.

———. 1984. *California landscape, origin, and evolution.* Berkeley: Univ. of Calif. Press.

Huber, N. King. 1987. *The geologic story of Yosemite National Park.* U.S. Geol. Survey Bull. 1595.

King, Clarence. 1872. *Mountaineering in the Sierra Nevada.* Boston: James R. Osgood and Co.

Matthes, François E. 1950. *The incomparable valley: A geological interpretation of the Yosemite.* Edited by F. Fryxell. Berkeley: Univ. of Calif. Press.

Muir, John. [1894] 1961. *The mountains of California.* Garden City, N.Y.: Doubleday and Co.

Russell, Israel C. 1889. *Quaternary history of Mono Valley, California.* U.S. Geol. Survey 8th Annual Report, 261–394.

Schoenherr, Allan A. 1992. *A natural history of California.* Berkeley: Univ. of Calif. Press.

Sharp, Robert P. 1988. *Living ice: Understanding glaciers and glaciation.* Cambridge: Cambridge Univ. Press.

INDEX

Boldface page numbers refer to figures and tables and their captions.

Design:	Barbara Jellow
Compositor:	Impressions Book and Journal Services, Inc.
Text:	10/13.5 Minion
Display:	Franklin Gothic Book and Semibold
Printer and Binder:	Edwards Brothers